Aquatic Dialogue Group:
Pesticide Risk Assessment and Mitigation

SETAC

A Professional Society for Environmental Scientists and Engineers and Related Disciplines Concerned with Environmental Quality

The Society of Environmental Toxicology and Chemistry (SETAC), with offices in North America and Europe, is a nonprofit, professional society that provides a forum for individuals and institutions engaged in the study of environmental problems, management and regulation of natural resources, education, research and development, and manufacturing and distribution.

Goals

- Promote research, education, and training in the environmental sciences
- Promote systematic application of all relevant scientific disciplines to the evaluation of chemical hazards
- Participate in scientific interpretation of issues concerned with hazard assessment and risk analysis
- Support development of ecologically acceptable practices and principles
- Provide a forum for communication among professionals in government, business, academia, and other segments of society involved in the use, protection, and management of our environment

Activities

- Annual meetings with study and workshop sessions, platform and poster papers, and achievement and merit awards
- Monthly scientific journal, *Environmental Toxicology and Chemistry*, SETAC newsletter, and special technical publications
- Funds for education and training through the SETAC Scholarship/Fellowship Program
- Chapter forums for the presentation of scientific data and for the interchange and study of information about local concerns
- Advice and counsel to technical and nontechnical persons through a number of standing and *ad hoc* committees

Membership

SETAC's growing membership includes more than 4,500 individuals from government, academia, business, and public-interest groups with technical backgrounds in chemistry, toxicology, biology, ecology, atmospheric sciences, health sciences, earth sciences, and engineering.

If you have training in these or related disciplines and are engaged in the study, use, or management of environmental resources, SETAC can fulfill your professional affiliation needs. Membership categories include Associate, Student, Senior Active, and Emeritus.

For more information, write SETAC, 1010 North 12th Avenue, Pensacola, Florida, 32501-3370, telephone 904-469-1500, or e-mail setac@setac.org.

Aquatic Dialogue Group: Pesticide Risk Assessment and Mitigation

Final report of the Aquatic Dialogue Group, May 1993 – May 1994

Publication sponsored by the
Society of Environmental Toxicology and Chemistry (SETAC)
and the SETAC Foundation for Environmental Education

Dialogue Group Members/Editors

James L. Baker, Iowa State University
Aldos C. Barefoot, E. I. duPont de Nemours & Co.
Larry E. Beasley, A. Duda and Sons, Inc.
Lawrence Burns, USEPA
Peter Caulkins, USEPA
Jerry Clark, National Fish & Wildlife Foundation
Richard L. Feulner, Ciba-Geigy Corporation
John P. Giesy, Michigan State University
Robert L. Graney, Miles, Inc.
Ray Griggs, Texas A & M University
Hank Jacoby, USEPA
Dennis Laskowski, DowElanco
Anthony Maciorowski, USEPA
Ellen Mihaich, Rhône-Poulenc
Henry Nelson, USEPA
Rodney Parrish, SETAC/SETAC Foundation
Richard E. Siefert, USEPA
Keith R. Solomon, University of Guelph
William van der Schalie, USEPA

Financial Contributors

American Crop Protection Association
United States Environmental Protection Agency

Published by SETAC Press
Pensacola, Florida

Cover by Michael Kenney Graphic Design & Advertising
Editing and typesetting by Wordsmiths Unlimited

Published by SETAC Press
SETAC Press is an imprint of the Society of Environmental Toxicology and Chemistry.

ISBN 1-880611-01-5 Paperback
ISBN 1-880611-05-8 Hardback
Library of Congress Catalog Card Number 96-67784
Paperback edition 1994
Printed in the United States of America
03 02 01 00 99 98 97 96   10 9 8 7 6 5 4 3 2

♾ The paper used in this publication meets the minimum requirements of the American National Standard for Information Sciences—Permanence of Paper for Printed Library Materials, ANSI Z39.48-1984.

Reference listing: Baker JL, Barefoot AC, Beasley LE, Burns LA, Caulkins PP, Clark JE, Feulner RL, Giesy JP, Graney RL, Griggs RH, Jacoby HM, Laskowski DA, Maciorowski AF, Mihaich EM, Nelson Jr HP, Parrish PR, Siefert RE, Solomon KR, van der Schalie WH, editors. 1994. Aquatic dialogue group: pesticide risk assessment and mitigation. Pensacola FL: SETAC Press.

CONTENTS

List of Figures

List of Tables

FOREWORD

The meetings of the Aquatic Dialogue Group and the preparation of this report were administered by the Society of Environmental Toxicology and Chemistry (SETAC) and the SETAC Foundation for Environmental Education.

SETAC is an international, professional society of 4,500 members founded in 1979 to provide a forum for individuals and institutions engaged in the study of environmental problems, the management and regulation of natural resources, education, research and development, and manufacturing and distributing. It is the only professional society that specifically brings together environmental scientists and engineers from academia, government, industry, and public-interest groups to provide research, education, and training in environmental problem solving. SETAC provides a forum through meetings, publications, and workshops for professionals involved in the use, protection, and management of our environment. The goals of SETAC are achieved through activities such as

- holding an annual scientific meeting consisting of symposiums and scientific paper and poster presentations on topics related to environmental toxicology and chemistry;
- publishing a monthly journal, *Environmental Toxicology and Chemistry*, bimonthly newsletters, *SETAC News* and *SETAC-Europe News*, and special publications;
- organizing and sponsoring regional groups to provide a forum for the presentation of scientific data and for the interchange and study of information of more local concern; and
- providing advice and counsel to technical and nontechnical people, groups, and institutions about scientific issues through a number of standing and *ad hoc* committees.

The SETAC Foundation for Environmental Education supports SETAC in all its activities. The purpose of the SETAC Foundation is to encourage, support, and promote education, charitable, and research endeavors that enhance a scientific approach to assessing the risks and benefits of chemicals in the environment. The goals of the Foundation are 1) to enhance public understanding of the risks and benefits associated with the chemicals used in everyday life and 2) to strengthen the technical basis for environmental risk assessment by supporting the purposes of SETAC.

Being associated with SETAC and its more than 4,500 members from around the world ensures a balanced view of environmental issues, provides accessibility to multidisciplinary technical expertise unmatched in any other professional society dealing with environmental issues, and establishes immediate linkage with a network of environmental professionals on a global basis.

The activity of the Aquatic Dialogue Group was funded by the American Crop Protection Association and USEPA, as well as by individual contributions to the SETAC Foundation's general fund.

PREFACE

The Society of Environmental Toxicology and Chemistry (SETAC) and the SETAC Foundation for Environmental Education are pleased to present this publication, the first from SETAC Press. It is the result of open and extended dialogue between the U.S. Environmental Protection Agency (USEPA) and the regulated community regarding aquatic risk assessment and mitigation options for pesticides. This publication is the product of a series of seven meetings of the Aquatic Risk Assessment and Mitigation Dialogue Group (Aquatic Dialogue Group) that occurred between May 1993 and March 1994.

The efforts of the Aquatic Dialogue Group Steering Committee and the SETAC/SETAC Foundation Office staff are gratefully acknowledged. Generous financial support from the American Crop Protection Association and USEPA is also acknowledged.

Rodney Parrish
Executive Director
SETAC/SETAC Foundation for Environmental Education
Pensacola, Florida
February 1996

ACKNOWLEDGMENTS

The editors would like to acknowledge the Aquatic Dialogue Group participants who contributed their time to provide valuable insights, ideas, and suggestions to advance the science, practice, and application of ecological risk assessment for pesticides in aquatic environments. Additional support was provided by Dr. Paul Hendley of Zeneca Ag Products, who reviewed the report and prepared the glossary; by Dr. Foster L. Mayer, Jr. of the USEPA Gulf Breeze Laboratory, who provided a detailed review of the report; and by the following group of individuals who also reviewed the report:

DuPont Agricultural Products, Wilmington, DE
Raymond J. Layton
Mark H. Russell
C. John Peter
Robert W. Freerksen
Joseph J. Dulka
Daniel M. Linn

Fawcett Consulting, Huxley, IA
Richard S. Fawcett

Iowa State University, Ames, IA
Jerry Hatfield
Steve Oberle

University of Florida, Gainesville, FL
Kenneth L. Campbell

University of Illinois, Urbana, IL
Loren E. Bode

University of Minnesota, St. Paul, MN
Jim Anderson

USDA-ARS, Coshocten, OH
Lloyd B. Owens

USDA-ARS, University of Nebraska, Lincoln, NE
J. F. Power

USDA-SCS, Washington, DC
David Schertz

Special recognition should be given to Linda Longsworth for providing excellent logistics and support during the workgroup meetings. In addition, we would like to acknowledge the editorial and typesetting support from Wordsmiths Unlimited's Mimi Meredith.

EXECUTIVE SUMMARY

Aquatic Dialogue Group: Pesticide Risk Assessment and Mitigation

In March 1992, a Task Force established within the United States Environmental Protection Agency (USEPA) Office of Pollution Prevention and Toxic Substances (OPPTS) was charged with reviewing and assessing ecological and environmental fate data requirements for registration and reregistration of pesticides. The Task Force focused on developing a process to meet congressionally mandated reregistration decisions for existing pesticides, although it was assumed a similar process would be appropriate for new pesticide registrations. The major aspects of the policy decisions reached by the Task Force are briefly paraphrased below:

- Establish risk-based priorities to allow protective decisions in a timely fashion.
- Base initial decisions primarily on laboratory studies with less dependence on terrestrial and aquatic field studies.
- Provide better integration of the risk assessment and risk management processes.
- Use risk mitigation to the extent feasible to achieve acceptable risk reduction.
- Develop the concept of continuous improvement and develop strategies to characterize long-term ecological risk with less uncertainty.

Following the OPPTS Task Force decisions, the Office of Pesticide Programs (OPP) developed an implementation plan to provide people within and outside USEPA a conceptual outline of the new process. Within USEPA and the regulated community, the implementation plan came to be known as the **new paradigm**.

Because of the importance of the new paradigm in pesticide regulatory decisions, it was felt that an open scientific dialogue regarding aquatic risk assessment and mitigation options would assist in the technical implementation of the new paradigm. A preliminary proposal was developed and an *ad hoc* steering committee was formed to establish the Aquatic Risk Assessment and Mitigation Dialogue Group (Aquatic Dialogue Group) under joint sponsorship by the American Crop Protection Association and USEPA. Representatives from USEPA (regulators and researchers), agrichemical companies, academia, and environmental and agricultural interest groups participated in the process. The proposed mission of the Aquatic Dialogue Group was to

- review pesticide risk assessment and risk mitigation for aquatic ecosystems within the OPP's integrated risk assessment paradigm, and
- offer recommendations for technical implementation and use of risk mitigation measures in regulatory risk assessment scenarios.

The Aquatic Dialogue Group met seven times between May 1993 and March 1994 to discuss these issues. The Group was organized to address the specific technical components of the new paradigm, and various workgroups were established to address each aspect. The two primary components of ecological risk assessment, effects and exposure, formed the basis for two of the workgroups. The exposure characterization workgroup focused on procedures, primarily simulation modeling, for predicting exposure in aquatic ecosystems. A risk characterization workgroup addressed the integration of the exposure assessments with effects information to develop predictive risk assessment approaches. The final workgroup critically reviewed available options for mitigating or reducing exposure to pesticides and discussed potential approaches for verifying the effectiveness of the mitigation in reducing risk.

The workgroups addressing the various aspects of the new paradigm covered an extensive array of technical issues and reached numerous conclusions and recommendations applicable to the paradigm's implementation. A review of some of the highlights follows.

EXPOSURE CHARACTERIZATION—The estimation of exposure of aquatic organisms to pesticides is a critical part of aquatic risk characterization. Exposure to pesticides generally occurs from two pathways: surface runoff and drift. Prior to the new paradigm, expected environmental concentrations (EECs) were represented as single values whose probabilities of occurrence or exceedance were not defined. It is recommended that such representations be used only in preliminary risk assessments and that refined risk assessments present exposure as distributions of EECs and a probability of exceeding a specific concentration.

Such probabilistic presentation of exposure data requires the use of computer simulation models, and the Aquatic Dialogue Group strongly supports the continued development and validation of such models. Numerous models are available, and at this point, the greatest limitation in applying such models in a probabilistic framework is generally the lack of model inputs. Model validation is still critically needed and will require additional field validation studies. The best field design for runoff model validation is small plots receiving simulated rainfall, although some model validation information can be obtained from studies focusing on larger areas (*i.e.*, watersheds).

The numerous technical recommendations cited in this report should be assigned to the Federal Insecticide, Fungicide and Rodenticide Act (FIFRA) Exposure Modeling Work Group (EMWG) for further discussion and resolution. Information about this workgroup can be obtained from personnel in the Surface Water Section of the Environmental Fate and Effects Division of the Office of Pesticide Programs or from the American Crop Protection Association.

EFFECTS CHARACTERIZATION—In addition to understanding exposure, a risk characterization requires an estimate of the effects that may occur at given concentrations. The Aquatic Dialogue Group reviewed current toxicity testing requirements for pesticide registration and, in general, found them to be adequate, although specific recommendations for improvement are provided in this report. It was suggested that acute and chronic testing of additional species may be useful if the risk characterization shifts to a probabilistic presentation of the effects data (as recommended below). It was also recommended that specific risk quotients be replaced by estimates of specific effect levels (*i.e.,* LC5). A modified tier system was outlined, identifying the triggers for requiring specific toxicity tests. It was also recommended that, for refined risk assessments, consideration should be given to more focused toxicity studies, which more accurately define or reflect the exposure profile in the environment.

RISK CHARACTERIZATION—Characterizing the potential risk of a pesticide requires a comparison of the exposure potential with some measure of potential effects. It is recommended that the final risk characterization be presented as a distribution of toxicity values, rather than a single point estimate, and be compared with the exposure probability distribution. The degree of overlap between the exposure probability curve and the effects curve can be used by scientists and risk managers to estimate the probability that a certain percentage of aquatic species will be adversely affected.

MITIGATING UNACCEPTABLE RISK—The foundation of the new paradigm is immediate mitigation to reduce unacceptable risk. **Risk mitigation** or **risk reduction** is defined here as actions to reduce or eliminate concentrations of pesticides that may affect aquatic habitats and their components. Mitigation should reflect the best practical source control technologies; it must be pragmatic, achievable, and enforceable; and it should be consistent with other federal, regional, state, or local resource protection and pollution prevention initiatives. In addition, any mitigation measure implemented must provide meaningful ecological risk reduction. Mitigation options for significant reductions in runoff and/or drift are currently available and are reviewed in detail in this report.

VERIFICATION OF MITIGATION EFFECTIVENESS—Depending upon the types, nature, and understanding of the identified risks and accepted mitigation actions, additional data may be required to verify that mitigation actually reduces risk to an acceptable level. For the most part, verification will focus on quantifying reductions in exposure. The Aquatic Dialogue Group recommends a three-phase process for verification:

Phase 1 The initial verification phase may be either qualitative or quantitative, should primarily rely on simulation models (if available), should be conducted using existing data sets, and should not usually involve additional laboratory or field data.

Phase 2 This phase focuses on field or laboratory studies to improve estimates of the success of mitigation in reducing exposure. Field work may be required; however, the focus of such work should be the validation of runoff model estimates. Additional data on drift should not generally be required since such information is being generated by the Spray Drift Task Force (SDTF).

Phase 3 The final phase may involve field investigations of the actual performance of mitigation and wider scale analysis of residues of the pesticide in the environment. On occasion, studies on the status and dynamics of specific organisms and ecosystems of concern may be appropriate.

RECOMMENDATIONS

Within each of this report's chapters, recommendations are provided which address specific topics. From these specific recommendations, eight major recommendations have been collated and represent the primary factors the Aquatic Dialogue Group believes are essential if the risk assessment process for pesticides is to progress within the new paradigm.

Recommendation 1

Integrated probabilistic risk assessment approaches that include both the probability of exposure and effects should be implemented within the OPP. Probabilistic presentation of risk provides risk managers with information on the level of risk and uncertainty associated with specific regulatory decisions.

Path Forward: Exposure characterization within OPP should continue to incorporate the concepts of probabilistic exposure assessment. This should be done in cooperation with academia, industry, and other government agencies via the FIFRA EMWG. For the effects component, a strategy and a time line need to be developed within USEPA, possibly in cooperation with members from the Aquatic Dialogue Group, to switch from single EEC quotient method approaches to probabilistic approaches. Such changes in the risk assessment process need to be documented, formalized, and approved through appropriate management and policy channels.

Recommendation 2

Improved capabilities for predictive risk assessments through tiered modeling and focused laboratory studies should be encouraged and, when conducted, should be included as part of the refined risk assessment. Model validation is critical and, under most conditions, should be considered the primary objective of any field exposure work.

Path Forward: When appropriate, toxicity tests that simulate actual field exposure conditions, such as pulsed dose experiments, should be considered.

Recommendation 3

Mitigation must provide meaningful ecological risk reduction, be pragmatic and achievable, and consider the need for timely decisions and cost-effective utilization of financial and human resources. In addition, mitigation should be consistent with other federal, regional, and state resource protection and pollution prevention initiatives.

Path Forward: Consideration of these factors should be incorporated into OPP policy, and any suggested mitigation measures should be compared against these factors prior to implementation.

Recommendation 4

The phased approach for evaluating the efficacy of mitigation measures, which is recommended and detailed in this report, should be adopted via the appropriate USEPA channels as the current working guideline. This framework must consider the need for timely, environmentally protective decisions.

Path Forward: Case studies should be prepared that document instances of effective mitigation under diverse use conditions.

Recommendation 5

Collaborative efforts among registrants, government, and academia should be encouraged when addressing mitigation verification issues. Such collaborative efforts could include cooperation to develop data bases on mitigation efficacy, model validation, and monitoring networks. Such efforts will enhance efficiency and help ensure a consistent and usable data base.

Path Forward: Both USEPA and registrants need to fulfill their roles if a coordinated and collective strategy for addressing procedures for verification of mitigation is going to work. USEPA must develop a policy that encourages the formation of workgroups to address common problems, and the industry must be willing to share data and resources. A policy or procedure that enhances the opportunity for such coordination should be developed.

Recommendation 6

The effectiveness of the concepts of the new paradigm in improving the ability of USEPA to implement timely and protective environmental decisions must be monitored and evaluated on a routine basis. This should include a technical evaluation of the effectiveness of specific mitigation approaches.

Path Forward: Every three years, starting in 1996, the successes and failures of the new paradigm should be reviewed. Such a review should include USEPA personnel and outside experts (academia, industry, and special interest groups).

Recommendation 7

The Aquatic Dialogue Group supports the general trend of USEPA toward watershed/holistic approaches for risk assessment and risk management. Such approaches necessitate the incorporation of "landscape factors," via probabilistic risk assessments, into risk management decisions.

Path Forward: State, regional, and national regulators and resource management agencies should emphasize watershed/holistic approaches.

Recommendation 8

Risk managers must focus on developing a better understanding of the scientific principles governing risk assessment. The ability to make appropriate risk management decisions requires a basic understanding of the scientific components of risk assessment.

Path Forward: Risk managers must develop a base-level understanding of the science of ecological risk assessment and the principles of probabilistic risk assessment and field validation recommended in this report.

INTRODUCTION

In March 1992, an Ecological Effects and Environmental Fate Task Force was established within the U.S. Environmental Protection Agency (USEPA) Office of Pollution Prevention and Toxic Substances (OPPTS). The OPPTS Task Force was charged with reviewing and assessing ecological and environmental fate data requirements for registration and reregistration of pesticides. Issues considered during the Task Force deliberations included resource requirements necessary to review data, the utility of the data in risk management and regulatory decision making, and the impact of the data requirements on meeting congressionally mandated deadlines for the reregistration of pesticides already in use. While the Task Force focused on developing a process to meet congressionally mandated reregistration decisions for existing pesticides, it was generally assumed that a similar process would be appropriate for new pesticide registrations.

The full text of the policy decisions released by the OPPTS Task Force in October 1992 is provided elsewhere (Appendix A), and its major points are briefly paraphrased below.

- Establish risk-based priorities to allow protective decisions in a timely fashion.
- Base decisions primarily on laboratory studies with less dependence on terrestrial and aquatic field studies.
- Provide better integration of the risk assessment and risk management processes.
- Use risk mitigation to the extent feasible to achieve acceptable risk reduction.
- Develop the concept of continuous improvement and develop strategies to characterize long-term ecological risk with less uncertainty.

The policy decisions reflect several USEPA programs and initiatives that have emerged over the last few years, including pollution prevention, pesticide use reduction, and integrated risk assessment and risk management strategies.

Following the OPPTS Task Force decisions, the Office of Pesticide Programs (OPP) convened an Ecological Effects and Environmental Fate Implementation Work Group. This second USEPA workgroup was charged with developing a plan to implement the OPPTS Task Force decisions. The OPP implementation plan was completed in August 1993 (Appendix B), and was developed to provide people within and outside the USEPA with a conceptual outline of the new process. Within the USEPA and the regulated community, the policy decisions and implementation plan came to be known as the **new paradigm**. Accordingly, this terminology has been retained throughout this document, and specifically refers to emerging pesticide registration and reregistration activities arising from the two USEPA workgroups.

Because of the importance of the new paradigm to the Environmental Fate and Effects Division of USEPA's Office of Pesticide Programs and the regulated community, preliminary discussions regarding the new paradigm were held with several agrichemical company technical representatives and the American Crop Protection Association in the fall of 1992. The Association and USEPA concurred that an open scientific dialogue regarding aquatic risk assessment and mitigation options would assist appropriate technical implementation of the new paradigm. Scientific dialogue was a logical extension of the implementation plan initially developed by OPP (Appendix B), which was largely a conceptual outline that was expected to evolve with time and experience. Implementation steps for the new paradigm will be further discussed in the Background section of this chapter.

A preliminary proposal was developed and an *ad hoc* steering committee was formed to establish the Aquatic Risk Assessment and Mitigation Dialogue Group (Aquatic Dialogue Group) under joint American Crop Protection Association and USEPA sponsorship. The Society of Environmental Toxicology and Chemistry (SETAC) Foundation for Environmental Education was selected to facilitate scientific dialogue between USEPA (regulators and researchers), agrichemical companies, academia, and environmental and agricultural interest groups (Table 1).

The proposed mission of the Aquatic Dialogue Group was to

- review pesticide risk assessment and risk reduction for aquatic ecosystems within the OPP's integrated risk assessment paradigm, and
- offer recommendations for technical implementation and use of risk mitigation measures in regulatory risk assessment scenarios.

The Aquatic Dialogue Group met seven times between May 1993 and March 1994 to discuss these issues and prepare a summary document of its proceedings. In the first meeting, the Group evaluated the scope of its mission and judged it necessary to review and evaluate the entire new paradigm. Specifically, the Group examined methods for assessing the efficacy of mitigation, and the group proposed and critiqued methods for assessing environmental exposure and ecological hazard of pesticides to aquatic systems. This document presents the Aquatic Dialogue Group's deliberations concerning aquatic risk assessment procedures, when aquatic mitigation would be necessary for agricultural use of pesticides, followed by mitigation methods that are currently available. Finally, methods for assessing the efficacy of mitigation actions to reduce or eliminate exposure and/or effects to aquatic resources are discussed.

TABLE 1. Aquatic Dialogue Group Participants

Dr. James L. Baker
Iowa State University
Dept. of Agric. & Biosystems Eng.
Ames, Iowa 50011
T 515-294-4025
F 515-294-2552

Dr. Aldos C. Barefoot
E.I. duPont de Nemours & Co.
Research Associate
Environmental Studies Section
Agricultural Products,
Experiment Station
Wilmington, DE 19880-0402
T 302-695-1583
F 302-695-4296

Dr. Larry E. Beasley
A. Duda and Sons, Inc.
Director of Crop Research
& Environmental Compliance
Box 257
Oviedo, FL 32765
T 407-365-2111
F 407-365-2010

Dr. Lawrence Burns
USEPA
ERL - Athens
960 College Station Road
Athens, GA 30605
T 706-546-3511
F 706-546-3252

Dr. Peter Caulkins
USEPA
Mail Code 7508W
401 M Street, SW
Washington, DC 20460
T 703-308-8000
F 703-308-8005

Dr. Jerry Clark
National Fish & Wildlife Foundation
1120 Connecticut Ave., NW.
Suite 900
Washington, DC 20006
T 202-857-0166
F 202-857-0162

Mr. Richard L. Feulner*
Ciba Plant Protection
Ciba-Geigy Corporation
Director, Regulatory Affairs
Box 18300
Greensboro, NC 27419-8300
T 910-632-2700
F 910-292-6374

Dr. John P. Giesy*
Michigan State University
Dept. of Fisheries & Wildlife
Pesticide Research Center
Institute for Environmental Toxicology
#13 Natural Resource Bldg.
East Lansing, MI 48824-1222
T 517-353-2000
F 517-432-1699

Dr. Robert L. Graney*
Miles, Inc.
Manager, Ecological Effects Group
17745 South Metcalf
Stilwell, KS 66085-9104
T 913-897-9132
F 913-897-5215

Mr. Ray Griggs
Texas A&M/Blackland
808 Blackland Road
Temple, TX 76502
T 817-770-6631
F 817-770-6561

Mr. Hank Jacoby*
USEPA
Mail Code 7507C
401 M Street, SW
Washington, DC 20460
T 703-305-7327
F 703-305-6309

Dr. Dennis Laskowski*
DowElanco
Senior Advisor
9410 Zionsville Rd., A-2/723
Indianapolis, IN 46268-1053
T 317-337-3470
F 317-337-3235

Dr. Anthony Maciorowski*
USEPA
Mail Code 7507C
401 M Street
Washington, DC 20460
T 703-305-5365
F 703-305-6309

Dr. Ellen Mihaich
Rhône-Poulenc
Toxicology
Box 12014
Research Triangle Park, NC
27709-2014
T 919-549-2025
F 919-549-2925

Dr. Henry Nelson
USEPA
Mail Code 7507C
401 M Street
Washington, DC 20460
T 703-305-7356
F 703-305-6309

Mr. Rodney Parrish* *ex officio*
Executive Director
SETAC/SETAC Foundation
1010 North 12th Avenue
Pensacola, FL 32501
T 904-469-9777
F 904-469-9778

Mr. Richard E. Siefert
USEPA
Ecosystems Response Branch
6201 Congdon Boulevard
Duluth, MN 55804
T 218-720-5552
F 218-720-5539

Dr. Keith R. Solomon
Centre for Toxicology
620 Gordon Street
Guelph, ON N1G 1Y4
CANADA
T 519-837-3320
F 519-837-3861

Dr. William van der Schalie
USEPA
Risk Assessment Forum
(RD-672)
401 M Street
Washington, DC 20460
T 202-260-4191
F 202-260-3955

*Steering Committee

BACKGROUND

The term **pesticide** includes any substance or mixture of substances intended for preventing, destroying, repelling, or mitigating any pest or intended for use as a plant regulator, defoliant, or desiccant. For simplicity, the term **pesticide** as used throughout this document represents pesticidal active ingredients and their toxicologically significant degradation products. The agricultural benefits of pesticides (insecticides, herbicides, fungicides, miticides, nematocides, etc.) are well known. Pesticides are partially responsible for the tremendous increases in user yields over the last few decades. It is realized that without continued pesticide use, input prices would increase while production per unit acre would decrease.

Although the agricultural benefits of many pesticides are well documented, pesticides are inherently toxic and may pose risks to nontarget organisms. As such, the USEPA evaluates a pesticide's potentially adverse effects, as well as its economic benefits, to human health and the environment before the pesticide is registered for use in the United States. In addition to the registration of new pesticides, currently registered pesticides are being evaluated in an extensive reregistration process mandated by Congress. The intent of reregistration is to reevaluate all registered pesticides by 1997 and subsequently to reevaluate existing registration on a regular basis (*e.g.*, every five years).

The overall data evaluation and decision-making responsibilities within the Office of Pesticide Programs are broadly divided between scientific and risk management divisions. The scientific divisions consist of the Health Effects Division (HED), the Environmental Fate and Effects Division (EFED), and the Biological Evaluation and Analysis Division (BEAD). These three divisions are responsible for human health risk assessments, environmental risk assessments, and agricultural benefit assessments, respectively. The science divisions review data from pesticide registrants and other sources to develop exposure, hazard, or economic profiles and risk assessments to support regulatory decision making. The risk management divisions include the Registration Division (RD) and the Special Review and Reregistration Division (SRRD). The lead divisions integrate the information provided by the science divisions and are ultimately responsible for final registration, reregistration, and other regulatory decisions (*e.g.*, granting registration, extending registrations, cancellation of specific pesticides or pesticide uses, etc.). Although the focus of this document is on aquatic ecological risk assessment and mitigation options largely centered in EFED, it must be emphasized that final regulatory decision making occurs in a broader multidisciplinary scientific, legal, and policy framework.

The Federal Insecticide, Fungicide and Rodenticide Act (FIFRA) requires environmental fate and effects studies that include data be submitted to the USEPA in support of the registration of a pesticide for a specific end-use. Under Section 3(2)(A) of FIFRA, the Administrator of USEPA is required to

> "... publish guidelines specifying the kinds of information which will be required to support the registration of a pesticide ...."

The suite of tests required for a specific pesticide registration or reregistration is dependent upon the intended use pattern of the pesticide, as well as its chemical, physical, and toxicological properties. Results of basic laboratory studies may trigger more complex studies in a tiered fashion and have often included field studies. Detailed presentation of specific test requirements and conditions that mandate when such studies are required is beyond the scope of this document. However, data requirements for pesticide registration are published through the USEPA's formal rule-making procedures in 40 Code of Federal Regulations Part 158, Office of Pesticide Programs guidelines, standard evaluation procedures, and technical guidance documents. Documents pertaining to environmental fate and effect data requirements are available through the National Technical Information Service (Springfield, Virginia) and are listed in Appendix C.

As traditionally determined in the Office of Pesticide Programs, **ecological risk** assessment is expressed as a function of ecotoxicological effects and environmental exposure. Aquatic **ecotoxicological effects** may be direct or indirect and have most often been measured through laboratory acute and chronic toxicity tests measuring LC50, EC50, or NOECs, or by observing field effects on aquatic organisms. **Environmental exposure** assessment consists of characterizing residues (measurement or model estimates of their levels) in the environment, the duration of exposure to nontarget organisms, and the types and kinds of nontarget biota that may be exposed to chemical residues of concern. **Ecological risk** is subsequently characterized by comparing the toxicological endpoints of interest with the environmental exposure to determine the ecological risk(s).

In the pesticide registration or reregistration process, ecological risk assessment procedures have entailed several discrete activities:

- The review and evaluation of ecotoxicological and fate and transport data submitted by registrants in support of the registration or reregistration
- The exposure evaluation includes the identification of environmentally significant residues, environmental compartments into which these residues may mitigate, and may include an estimate of residue concentrations in the environmental compartments of concern for a given time period after application as a function of time and location after application.
- Establishment of the ecological endpoints of concern for the pesticide and its use(s) and the toxicity to nontarget organisms and their habitat based upon an evaluation of submitted data
- Calculation of risk quotients as defined by exposure (as defined by estimated environmental concentration [EEC]) divided by hazard (acute and chronic effects) based upon ecological toxicity data, pesticide use data, fate and transport data, and estimates of environmental exposure
- A number of ecological levels of concern (LOCs) may be used in OPP regulatory decision making (Table 2). LOCs consist of quotient comparisons of estimated hazard (acute and chronic toxicity for fish, invertebrates, and plants) with the EEC. In addition, values may be derived for endangered species and supplemented with field data or incident data for the risk characterization.

TABLE 2. Ecological Levels of Concern Currently Used in OPP Regulatory Decision Making (The LOCs are a quotient comparison of a measured hazard with the EEC.)

The LOCs for acute aquatic effects:	If the EEC/LC50 ≥ 0.5, the potential acute risk to aquatic organisms is high and may warrant regulatory action in addition to restricted use classification.
	If 0.5 > EEC/LC50 ≥ 0.1, then the pesticide is considered for restricted use classification.
	If EEC/LC50 < 0.1, the potential acute risk to aquatic organisms is low, and no additional regulatory action is warranted.
The LOCs for chronic aquatic effects:	If the EEC ≥ LOEC, the potential chronic risk to aquatic organisms is high and may warrant regulatory action.
The LOCs for acute effects to endangered species:	If the EEC/LC10 ≥ 0.1 (when a slope is available) or 1/20 LC50 (when no slope is available), the potential acute risk to endangered aquatic species is high and may warrant consultation with the Office of Endangered Species (OES).
	If EEC/LC10 < 0.1 (when a slope is available) or EEC/LC50 < 0.2 (when no slope is available), the potential acute risk to endangered species is low, and no consultation with OES is necessary.
The LOCs for chronic effects to endangered species:	If the EEC ≥ LOEC, the potential chronic risk to endangered aquatic species is high and may warrant consultation with OES.
	If the EEC < LOEC, the potential chronic risk to endangered aquatic species is low, and no consultation with OES is necessary.

LOEC - Lowest Observed Effects Concentration

- If LOCs are exceeded, a refined risk assessment may be conducted. Refinements may address the toxicological or exposure component of the risk assessment. Once the core set of studies is completed, additional exposure or toxicity information that better predicts potential field effects may be considered (*e.g.*, pulsed exposure experiments reflecting environmental exposure). Exposure, hazard, and risk estimates may also be refined by considering the extent of pesticide use, incident reports, kinds of species exposed, and quality of treated habitats. A comparative analysis of alternative pesticides for the use in question may be conducted. A comprehensive analysis is next performed in which the duration, extent, and frequency of the risk from the pesticide use are characterized. The information is summarized in an ecological risk conclusion and communicated to the risk management divisions to assist regulatory decision making.

To this point in the background discussion, we have centered on risk assessment methods and procedures that were routinely employed by OPP prior to the development of the new paradigm. It should be emphasized that none of the risk assessment methods and procedures discussed above are eliminated under the new paradigm. However, the new paradigm represents a fundamental shift in thinking that focuses attention on making timely regulatory decisions through better integration of risk assessment, risk management, and risk mitigation.

The new paradigm is perhaps best viewed as the application of regulatory science within an informed risk management context. This view is supported by emerging risk-based approaches to environmental regulations (Thomas, 1987; Science Advisory Board 1990a, b, c, d), which promote increased integration of societal values, science, and risk mitigation practices for environmental decision making. The integrated decision-making process involves three interactive phases.

1. **Risk assessment** is a science-based process that consists of hazard identification and exposure assessment, and that ultimately integrates hazard and exposure to characterize risk.
2. **Risk mitigation** involves remediation or mitigation measures to reduce or eliminate source contamination and adverse environmental impact.
3. **Risk management** is a policy-based activity that defines risk assessment questions and endpoints to protect human health and the environment. It takes the scientific risk assessment and incorporates social, economic, political, and legal factors that impinge or influence the final decision and selects regulatory actions.

One controversial aspect of the new paradigm is reduced emphasis on field studies to initially characterize ecological risk. Reduced reliance on field studies has been strongly criticized by ecologists and ecotoxicologists who believe in the value of such studies. In the recent past, aquatic field studies or simulated field studies were required with the objective of rebutting the presumption of risk when LOCs identified through laboratory studies for a particular pesticide and use pattern were exceeded. The new paradigm clearly de-emphasizes long-term field studies to rebut a presumption of risk. However, it also assumes that LOC exceedances derived from laboratory studies are sufficient to institute mitigation actions to reduce risk, and that monitoring and/or field studies may be necessary to verify that exposure or effects have been mitigated.

Risk reduction or risk mitigation activities are defined as actions to reduce or eliminate concentrations of pesticides in the environment that may affect human health, or aquatic and terrestrial habitats and their component biota. Risk mitigation can involve numerous regulatory activities, from labeling restrictions to cancellation of specific uses. Mitigation measures should reflect the best practical nonpoint control technologies, processes, or other measures that will provide the greatest achievable reduction in the exposure of nontarget aquatic organisms and habitats to pesticides. Although it is believed that mitigation measures can reduce environmental exposure and thus potentially adverse effects, there has been little systematic identification or evaluation of appropriate mitigation measures.

The new paradigm places less reliance on rebutting the presumption of risk with field studies and promotes the concept of risk reduction when LOC exceedances are identified through the risk assessment process. The focus of the new paradigm is to promote environmentally protective decisions that consider mitigation actions to reduce the off-site movement of pesticides and therefore reduce risk to nontarget organisms. There are provisions for more sophisticated assessment procedures that allow for probabilistic estimates of levels of concern. The new paradigm also shifts pesticide risk assessment away from a purely prospective process by promoting mitigation and retrospective evaluations through monitoring.

RISK IDENTIFICATION AND MITIGATION UNDER THE NEW PARADIGM

As described earlier, basic data requirements and general risk assessment procedures routinely employed by OPP before the development of the new paradigm are largely unchanged. However, the new process entails substantially greater interaction and communication between scientists involved in the risk assessment phase of the process and risk managers responsible for final decision making. The implementation steps, roles and responsibilities, and communication steps within the various OPP divisions are detailed in Appendix B and briefly summarized in Figure 1.

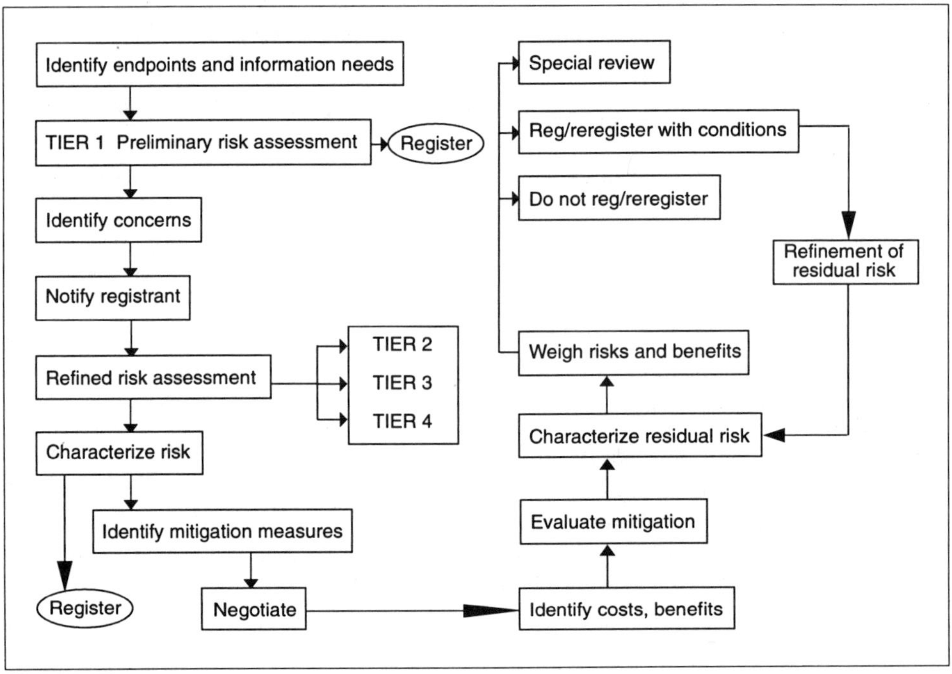

Figure 1. Overall approach to risk assessment and mitigation

The four levels for refining risk assessment identified in Figure 1 need to be put into the context of ongoing regulatory programs. Under the reregistration program, the USEPA has legislatively mandated time frames for making decisions. The USEPA is committed to making the best decisions possible with the information available at the time. Delaying regulatory decisions while further assessment refinements are undertaken is not a viable option. The responsibility belongs with the registrant to determine the appropriate level of refinement and to provide the information in a timely manner consistent with established Reregistration Eligibility Decisions (RED) dates. If such refinements are provided, the USEPA should ensure this additional information is used when evaluating mitigation options. The registrants are strongly encouraged by the USEPA to undertake these refinements since the result would be a more precise identification of areas where mitigation would likely be warranted.

Although a salient feature of the new paradigm is increased interaction between risk assessors and risk managers, the new paradigm clearly delineates between the risk assessment and risk management roles of the process, based on recommendations of USEPA's Risk Forum (USEPA, 1992b). The prospective risk assessment portion of the new paradigm is shown on the left side of Figure 1. The prospective risk assessment remains essentially a scientific exercise and may be iterative. That is, the data base supporting a current pesticide use registration will be periodically reviewed and updated to ensure it meets appropriate scientific requirements and regulatory policies.

When a pesticide undergoes evaluation for reregistration or registration, scientists review and evaluate the data base in a comprehensive manner. In registration, the supporting data are reviewed and evaluated to ensure they meet the standards established for carrying out the studies in the data base. The data base is evaluated and integrated such that residue levels, routes of dissipation, and the formation and persistence of significant environmental degradates in various environmental compartments are elucidated. This information is evaluated in conjunction with toxicity test endpoints determined by a suite of required studies and available incident data to ascertain whether predefined (Table 2) LOCs have been exceeded for the compound and use under evaluation.

The preliminary risk assessment will generally employ conservative hazard estimates and use scenarios resulting in maximum exposure, unless sufficient data are initially available to refine the risk assessment during the initial attempt. If the preliminary risk assessment yields no LOC exceedances, the risk assessment will be finalized and sent to the appropriate risk management division (Registration Division [RD] or Special Review and Reregistration Division [SRRD]) for incorporation into the final multidisciplinary decision process.

If LOC exceedances occur in the preliminary risk assessment, a conclusion that an unacceptable risk may result from the proposed or registered use(s) of pesticides is immediately passed to the appropriate risk management division within the Office of Pesticide Programs. If sufficient data to better characterize exposure or hazard are available, the risk estimate may be recalculated to verify or refute the preliminary risk conclusion. When LOC exceedances are refuted by the refined risk assessment, a revised risk conclusion will be transmitted to the risk management division.

When the refined risk assessment verifies the preliminary risk assessment or when there are insufficient data to conduct a refined risk assessment, a risk conclusion stating the nature of the exceedance and proposed preliminary mitigation options will be prepared for transmittal to the risk management division and the Biological Evaluation and Analysis Division (BEAD). BEAD will provide a preliminary evaluation of the proposed mitigation options to determine their economic and agricultural feasibility relative to user-specific agricultural practices. The risk management division will then notify the registrant of the risk conclusion and preliminary feasibility analysis of the mitigation proposals.

Once the prospective risk assessment (preliminary or refined) is sent to the appropriate risk management division (RD or SRRD), the risk management phase of the process (shown on the right side of Figure 1) begins. The risk management division has the responsibility of negotiating with the registrants and coordinating interactions between scientists and risk managers to identify the risks and benefits of a pesticide use, possible data needs to decrease uncertainty, mitigation and monitoring options (where appropriate), the relative effect of mitigation on reducing risk, and a final risk/benefit determination.

The purpose of notification is to ensure everyone involved in the decision is actively involved in the process of identification of appropriate risk concerns, agricultural benefits of the pesticide use, and risk mitigation and monitoring measures. This process may include scientists and economists from OPP and industry, as well as user groups. Once the risk reduction measures have been established, they will be considered in the final decision process, which may entail pesticide label changes necessary to reduce the risk, registration with conditions, as well as a decision to cancel a use or deny a registration.

The ideal product of the overall risk assessment, risk reduction, and risk management process is a set of mitigation actions to reduce environmental risk to an acceptable level. As the new paradigm is implemented in regulatory decision making, data to support the effectiveness of the mitigation will often be scant. To ensure the effectiveness of the mitigation steps, the USEPA may require some sort of verification data.

Once mitigation measures have been identified and implemented, post-reregistration or registration evaluation of the efficacy of the mitigation may be required. Quantifiable verification of effectiveness of the mitigation may take several years. If verification information is required, the additional data would be submitted within an agreed-upon period. The verification data would then be reviewed, evaluated, and incorporated into the USEPA's data base for consideration in the next major review cycle for the chemical.

SELECTED REFERENCES

Science Advisory Board, USEPA. 1990a. Reducing risk: Setting priorities and strategies for environmental protection. SAB-EC-90-021. Washington, DC.

Science Advisory Board, USEPA. 1990b. The report of the Ecology and Welfare Subcommittee: Relative risk reduction project, reducing risk. Appendix A. SAB-EC-90-021A. Washington, DC.

Science Advisory Board, USEPA. 1990c. The report of Human Health Subcommittee: Relative risk reduction project, reducing risk. Appendix B. SAB-EC-90-021B. Washington, DC.

Science Advisory Board, USEPA. 1990d. The report of the Strategic Options Subcommittee: Relative risk reduction project, reducing risk. Appendix C. SAB-EC-90-021C. Washington, DC.

Thomas, L. M.. 1987. Environmental decision-making today. *Environmental Protection Agency Journal* 13:2-5.

USEPA. 1992a. *Framework for Ecological Risk Assessment.* EPA/600/R-92/001.

USEPA. 1992b. Memorandum from Linda Fisher, Assistant Administrator for Prevention, Pesticides and Toxic Substances, to Douglas Campt, Director of the Office of Pesticide Programs, entitled *Decisions on the Ecological and Environmental Fate Data Requirements Task Force,* October 29, 1992.

Chapter 1: OVERVIEW

The final report of the Aquatic Risk Assessment and Mitigation Dialogue Group (Aquatic Dialogue Group) is divided into two main topics. The first addresses procedures for conducting risk assessments and focuses on potential improvements in the process. Many of the concepts discussed are directly applicable within the refined risk assessment component of the **new paradigm** (see Introduction). The second discusses the options available for mitigating risk and outlines a process for evaluating the effectiveness of such mitigation measures. This summary provides a brief overview of the main points discussed in this report.

1.1 RISK ASSESSMENT

The risk assessment process for assessing levels of concern (LOCs), as well as the needs for mitigation that are recommended in this report, is based on a probabilistic assessment of exposure effects on aquatic organisms. This process makes better use of all the data developed for registration and allows for flexible integration with risk mitigation and management options. The proposed process is based on the current data requirements for registration, but it suggests somewhat different triggers and may under specific circumstances require more toxicological data. More extensive modeling of expected environmental concentrations (EECs) would be needed as well. It is proposed that the process be tiered to reduce unnecessary evaluation of pesticides that pose minimal threat to the aquatic environment. Early stages of the tiers can be implemented on the basis of current registration requirements, while later tiers will require more information than is currently available. The approach used was adapted from the *Framework for Ecological Risk Assessment* developed by the U.S. Environmental Protection Agency (USEPA, 1992).

1.1.1 Characterizing Exposure

The estimation of pesticide exposure to aquatic organisms is an important part of the aquatic risk assessment process. Before the new paradigm, EECs were developed in a variety of ways. All of these estimates represent single EECs whose quantitative likelihood of occurrence is unknown because they have not been placed in the probabilistic framework that is necessary to capture the high spatial and temporal variation of actual pesticide concentrations in the environment. We recommend that the new process use only these EECs in the preliminary risk assessment.

To be consistent with the new paradigm, emphasis should be placed on computer modeling to generate probabilistic distributions of EECs rather than delaying a risk assessment while waiting for monitoring estimates of environmental concentrations. In cases where acceptable monitoring data are already available, they will be used in the risk assessment. Single EECs should be replaced with temporal and spatial distributions of EECs that relate directly to the quantitative likelihood of their occurrence so that degrees of risk to the aquatic community can be quantitatively assessed and tied to mitigation or risk management if necessary. Such probabilistic EEC distributions for one to multiple sites over multiple years can be used to estimate where and how frequently the maximum or average concentrations of pesticides will exceed any given threshold for acute or chronic toxicity to aquatic organisms. The modeling approach can be further extended to identify specific market uses and environmental factors that result in high EECs, thus further aiding the management process.

There is currently a lack of complete understanding of interactions among many competing environmental fate processes involved in pesticidal degradation and dissipation, and as a result, it is sometimes difficult to construct appropriate models. Conceptually, however, it is generally possible to derive plausible models of environmental fate even though some may require additional validation. The conclusion at this time is that there are a number of adequate environmental fate models that can be used directly or adapted for use with probability-based parameter estimates, although the parameter estimates are often lacking. Thus, generating probabilistic distributions of EECs is generally more limited by available input data than by the availability of models.

1.1.2 Characterizing Effects

Effects are characterized primarily by using single-species laboratory toxicity tests. These data are initially used in a preliminary risk assessment to obtain point estimates of effects; for probabilistic estimates, a species sensitivity distribution is obtained. At a minimum, data from at least eight acute studies should be available to construct a species distribution curve. The species tested should include at least one fish, one invertebrate, and one plant. In addition, it is suggested that one additional organism be chosen, using the most sensitive group as identified in the suite of acute toxicity tests.

In order to get the best data and gain the most useful information for the risk assessor and the risk manager, it is important to be able to focus resources where they are most needed. In addition to the studies needed to determine the hazard probability distribution for the refined risk assessment, more focused studies could be performed in order to better define the toxicity profile in the environment. While these proactive tests would not serve to replace the required guideline studies, they should be considered during the risk assessment and risk management processes in order to better understand the actual risk to the aquatic environment. An example of how additional information could be beneficial relates to use pattern. If the pesticide is used once a week for a number of weeks, then, in addition to the standard acute toxicity test, it would be appropriate to utilize a pulse-dosed system to mimic the incremental inputs of the pesticide over time.

1.1.3 Risk Characterization

For a probabilistic characterization of risk, distributions of exposure and toxicity values are compared rather than the point estimates used in the quotient method. The government of the Netherlands presently uses a distributional method for evaluating species sensitivity data, and a similar approach is used in the United States in the calculation of water quality criteria. A major advantage of this approach is that it uses all relevant single-species toxicity data, and when combined with exposure distributions, allows quantitative estimations of risks to aquatic organisms. The degree of overlap between the exposure curve and the effects curve can be used to estimate the probability that a certain percentage of aquatic species may be adversely affected. In using overlaps of distributions, there is a tacit assumption that protecting a certain percentage of species for a certain proportion of occasions will also preserve ecosystem structure and function. Given the natural variability between sites within ecosystems (and even between ecosystems), this assumption will probably hold. Additional research may be required to validate this assumption.

There are some limitations to this approach. For example, the choice of protection level (*e.g.*, 90% of species) may not be socially acceptable. Some may view 90% as being overprotective, while others may find that degree of risk unacceptable, especially if the 10% of potentially affected species includes endangered species or other organisms of high ecological, commercial, or recreational significance. In addition, risks from persistent bioaccumulative chemicals to species at the top of the food chain may not be sufficiently considered by this approach alone, and these effects should be identified and appropriate mitigation measures taken.

1.1.4 The Tiered Approach

Currently, several triggers are used to proceed from acute to chronic testing in aquatic organisms. Some of these triggers may be overly conservative. Overall, the triggers for acute testing are acceptable. The following triggers are proposed for chronic tests:

- The pesticide is intended for use such that its presence in water is likely to be continuous or recurrent,
- the EEC $\geq$0.01 of the LC50 or EC50 based on concentration-response slope, and
- the half-life in water is greater than 4 days or the physicochemical properties indicate cumulative effects.

Other tests such as the full life-cycle studies on fish could be triggered by indications of reproductive effects in other organisms, from structure-activity assessments, or from the results of mammalian or avian tests carried out in response to other sections of the risk assessment process.

The current use of a fraction of the LC50 as the trigger for an LOC is judged inappropriate on the basis of knowledge of the concentration-response data and the analysis of currently used registration data. When appropriate and when the data are available, we recommend the use of a point-estimate of a low-response concentration calculated from the slope of the line such as the LC5. This better takes into account the potentially adverse, low-concentration effects of some pesticides. These data may be available and, if so, low-concentration effects can be calculated from the results of the tests (*e.g.*, Mayer *et al.* 1994; Sun *et al.*, 1994) currently required for registration.

The process of risk assessment should be tiered to reduce the need for unnecessary assessments and delays in the regulatory process (Figure 1-1). The first stage is a preliminary assessment designed to allow a rapid assessment of those pesticides that do not present any significant risk to the aquatic environment. By design, this step is conservative. The preliminary risk assessment is followed by a series of progressively more refined risk assessments in which more detailed information is used to again isolate those pesticides that do not present a significant hazard. Pesticides or specific uses of pesticides can then be identified for mitigation measures.

1.2 RISK MITIGATION AND VERIFICATION

An important component of the new paradigm is reducing exposure to mitigate risks. Depending on the types, nature, and understanding of the identified risks and accepted mitigation actions, additional data to verify that the mitigation actually reduces risk to an acceptable level may be required.

1.2.1 Risk Mitigation Options

Mitigation activities are defined here as actions to reduce the exposure of aquatic habitats to pesticides. Risk mitigation can involve numerous regulatory activities, ranging from label restrictions to cancellation of specific uses. Mitigation measures should reflect the best practical nonpoint control technologies, processes, or other measures that will provide the greatest achievable reduction in the exposure of nontarget aquatic organisms to pesticides.

Practicality on the part of the regulator, the registrant, and the grower must be considered when evaluating the need for mitigation. Requiring mitigation based on preliminary or faulty risk characterizations can create undue burden and cost for the user. Often, decisions of both the regulator to require mitigation and the grower to implement the mitigation require the use of "best professional judgment."

Regardless of the type of mitigation proposed for a specific pesticide product, the mitigation measure should exhibit several practical attributes:

- Mitigation options should be possible within the time frame allotted by the regulatory process. The mitigation measure should balance the desired risk reduction with the resources required for implementation.
- Mitigation should provide meaningful ecological risk reduction.
- Mitigation measures should be consistent with other federal, regional, state, and local resource protection and pollution prevention initiatives.
- Mitigation measures should be pragmatic, achievable, and enforceable.

When appropriate, the cluster concept should be considered when evaluating mitigation options and the effectiveness of mitigation. Clustering can be conducted based on use groups (*i.e.*, corn insecticides) and/or specific pesticide properties. Such an approach will reduce resource expenditures and minimize confusion in the user community regarding the specific mitigation practices required.

The first and foremost option for mitigation is optimization of the label, actually reducing the amount of product applied to the minimum required to control the pest. After the label is optimized, mitigation should attempt to minimize off-target movement of the pesticide. Numerous options are available for reducing exposure from either spray drift or surface runoff.

TIER 1 (DETERMINISTIC)

EEC ESTIMATION	RISK ASSESSMENT	DATA REQUIRED
Single-event EEC based on a high-exposure scenario (direct overspray of a pond 2 m deep or 1-10% runoff from a 10 ha field into a 1 ha pond)	Is the EEC < the LC5 for the most sensitive species? And, where chronic toxicity data are available, is the EEC < the chronic no observed effect concentration (NOEC) for the most sensitive species?	Acute toxicity and slope data for three species of freshwater (and saltwater) aquatic animals (as currently required), **and** one species of aquatic plant and **(if required)** chronic studies on four species of aquatic organisms

If "yes," no further assessment necessary. If "no," go to Tier 2.

TIER 2 (PROBABILISTIC)

EEC ESTIMATION	RISK ASSESSMENT	DATA REQUIRED
Reasonable high-exposure EECs modeled from a small number of soil/climate scenarios, each representing the upper 10th percentile of runoff yield (sediment and water) for 10-year return for each market	Identify each market for which the upper 10th percentile annual maximum EEC (for the duration of interest) for its reasonable high-exposure soil/climate scenario is < the lower 10th percentile of the distribution of the LC5 estimates from acute tests. Or, is this EEC < the lower 10th percentile distribution of the NOECs from chronic tests?	Additional acute toxicity tests on four species selected from the group of organisms (fish, algae, etc.) observed to be most sensitive in Tier 1 data. **If triggered** (by other criteria such as persistence), chronic studies on four species of aquatic organisms

If "yes," no further assessment necessary. If "no," go to Tier 3 or to mitigation.

TIER 3

EEC ESTIMATION	RISK ASSESSMENT	DATA REQUIRED
More scenarios to define geographical as well as climate-driven EECs (so problem areas within each market that fail Tier 2 can be identified) and to assess mitigation options	Identify each of the soil/climate scenarios within each market for which the upper 10th percentile annual maximum EEC (for the duration of interest) < the lower 10th percentile of the distribution of the LC5 estimates from acute tests. Or, is this EEC < the lower 10th percentile distribution of the NOECs from chronic tests?	No additional toxicity data required.

If "yes," no further assessment necessary. If "no," go to Tier 4 or to mitigation.

TIER 4

EEC ESTIMATION	RISK ASSESSMENT	DATA REQUIRED
Based on site-specific EECs (pulsed exposures) or landscape modeling (to assess position of treated areas in relation to surface waters or percentages of land used for a particular chemical)	As for Tier 3, or use more realistic toxicity tests	Regulator or registrant may require or furnish additional (field-derived) toxicity data that are more realistic, or special tests for pulse exposures or compounds with very short half-lives may be used to obtain a better estimate of the toxicity distribution.

If "yes," no further assessment necessary. If "no," mitigate.

Note: *The rationale for selection and use of the LC5 in the proposed tiered risk assessment process is provided in Chapter 2, Section 2.2.5.*

Figure 1-1. Proposed tiered risk assessment process

The details of specific mitigation measures are discussed in Chapter 4. A list of the available options appears in Tables 4-1, 4-2, and 4-3. Table 4-1 outlines generic mitigation approaches; Table 4-2 addresses options for mitigating runoff; and Table 4-3 lists measures useful for reducing or minimizing spray drift.

It is important to realize that mitigation options outlined in the tables are not always additive. They are discussed individually, and in some cases, a range of expected reductions in off-site transport is provided. However, the actual reduction in exposure cannot be estimated until all the options for a specific scenario are considered together.

1.2.2 Verification of Mitigation Effectiveness

When a pesticide has raised concern about potential ecological damage, proposed mitigation measures may be negotiated between the registrant and the USEPA, with appropriate input from affected user groups and state agencies. Given agreement on specific measures, it will be important to verify that the original concern has in fact been alleviated, *i.e.*, risk has been reliably reduced to acceptable levels. Verifying the success of mitigation measures in reducing risk can be thought of as a series of steps of increasing complexity and cost—steps to be taken only if the risk assessment process so requires. Each component of the risk assessment has an associated uncertainty, and the mitigation verification process will be driven primarily by the need to reduce these uncertainties. The phases are outlined below:

Phase 1. Phase I of verification is based upon statistical inference, simulation modeling, and expert judgment, using available data. If risks can be demonstrated to be mitigated to low levels with high certainty, then mitigation will have been "verified" by definition, and no additional action may be required. The reliability of this process flows from its use of the very same inference tools that have been used to identify the existence of significant risk and originally set in motion the mitigation process.

Given a set of appropriate mitigation proposals, three important activities can be identified:

1. use of simulation models to establish mitigation effectiveness,
2. use of best professional judgment and inference in evaluating mitigation effectiveness, and
3. evaluation of acceptability and practicality in the agricultural community.

Phase 2. When ecological concerns have not been alleviated by evaluation of the proposed verification methods in Phase 1, then a number of alternatives may be undertaken, including these four:

1. proposal of additional mitigation measures and return to Phase 1 risk assessments
2. efforts to achieve greater user acceptance of mitigation if data from initial evaluations indicate that users are simply not complying with the measures
3. site-specific field studies that are appropriate for model improvement and quantify the risk reduction achieved by each individual mitigation

4. site-specific investigations of the combined efficacy of the currently achieved practice

When modeling in Phase 1 is not adequate to evaluate mitigation, then experimental field trials may be required to investigate specific mitigation measures (*i.e.*, points 3 and 4 above). In such cases, studies simulating field conditions should be used to further develop the predictive capabilities of models. Measuring off-site transport can be linked with modeling to determine the effect of specific mitigation measures and to allow a prediction of mitigation effectiveness.

When mitigation is designed to reduce drift, the Spray Drift Task Force (SDTF) data base and associated models may provide an adequate basis for verifying the effectiveness of the mitigation. The landscape factors that may affect exposure (*e.g.*, wind speed) can be included in exposure models for evaluating mitigation or for refining the risk assessment.

Evaluating specific measures to reduce runoff can be done best through small-scale (0.1- to 0.2-acre) plots. These comparative studies examine the effect of factors such as buffer zones, application methods, formulations, or conservation practices. Such studies simulate field conditions by using well-characterized plots in which water, soil, and chemical transport are controlled through rainfall timing, intensity, and duration. The studies lend themselves to model validation, as well as to the development of reduction factors for minimizing transport off-site.

Phase 3. In unusual circumstances, the effect of a specific pesticide on a particular ecosystem will require residue and/or effects monitoring to confirm cause-effect relationships or to conduct a final confirmatory risk assessment. These studies will be required only when (1) there is a potentially high probability of an effect, (2) a biological effect has been demonstrated (*e.g.*, through incident reports), and/or (3) mitigation measures cannot be imposed without a significant reduction of the pesticide's benefits.

In Phase 3, studies can be designed to look at exposure and/or effects.

Exposure. General residue surveillance monitoring may be appropriate when it is necessary to determine actual exposure to a widely used chemical within a watershed or on a larger scale. Such studies should be required only when, after model validation, predicted high risks persist because of uncertainty in the realized performance of mitigation practices. Surface water and sediments may be monitored at specific locations within the watershed, but sampling would not be linked to specific sites, applications, or mitigation measures. Such studies inherently include the influence of all landscape factors on exposure, and thus residue measurements represent actual transport to the aquatic environment.

Effects. When considering the need for biological field studies to evaluate the efficacy of mitigation, the conceptual basis of predictive versus retrospective risk assessments should be separated. Retrospective risk assessments assume there is an identified problem (*e.g.*, fish kills) that specific field studies can be designed to evaluate and address. However, if specific events are not known to occur, then designing retrospective field studies to "prove" that such effects do not occur is inappropriate. For example, if risk managers, based on input and guidance from scientists, decide that it is important to determine the secondary effects of a chemical or that due to the pesticide's properties other specific information outside the standard array of tests would significantly improve the risk evaluation of the chemical, then additional work may well be justified. In such cases, studies should be conducted in experimental systems to provide predictive information.

Although the exposure and toxicological components of predictive risk assessments are sound, field verifications of prospective risk assessments for pesticides have seldom been conducted. Regardless, field biological monitoring studies for validation of the risk assessment should never be substituted for a poorly defined risk assessment. If biological monitoring studies are required, the need and use of the required data must be clearly articulated. Problems associated with high variability, lack of adequate reference sites, and difficulty in establishing cause-effect relationships must be defined before a specific biological monitoring requirement is considered. Other stressors (*i.e.*, habitat modification, soil erosion, nutrient additions, and other human activities) can also impact aquatic populations and must be appropriately addressed in the experimental designs. If there are specific uncertainties in the predictive risk assessment, laboratory or simulated field studies to improve the risk assessment should be considered before biomonitoring studies are required.

Improved incident reporting represents a form of "biological monitoring" that may be useful in evaluating mitigation effectiveness. For example, if recurrent fish kills are a primary factor identifying major concerns for a specific chemical use and mitigation is implemented to reduce exposure and eliminate such incidents, then it may be appropriate to evaluate whether fish kills are in fact stopped.

Although it has not been applied within the Federal Insecticide, Fungicide and Rodenticide Act (FIFRA) regulatory arena, field biological monitoring has been utilized in state and federal water quality investigations. Experience gained by other regulatory agencies in the use of biological monitoring should be considered when developing or evaluating the applicability of such studies to FIFRA regulatory needs. Studies must be designed that are known to provide scientists and risk managers with the information necessary to address specific regulatory questions or concerns.

1.2.3 Evaluating Long-Term Impacts of Pesticides

The development of procedures for evaluating long-term impacts of pesticides was a specific need identified by the Assistant Administrator's Ecological Effects and Environmental Fate Task Force (Appendix A). A variety of biomonitoring tools exists for such evaluations (*e.g.*, rapid bioassessment protocols, biological indices). These biomonitoring tools (like models) have not, however, been convincingly validated or specifically adapted for evaluating individual chemicals in aquatic habitats associated with agroecosystems. Validation and development work is required.

It is important to develop field data by establishing the relative condition of aquatic ecosystems in agricultural areas. When it is possible to link field effects to individual chemicals, such assessments should be done. However, linking long-term evaluations of ecosystem integrity with chemical-specific assessments, or with evaluating mitigation effectiveness for specific chemicals, would be very difficult. Concerns relative to long-term impacts should be integrated into existing or planned programs within USEPA or other federal and state agencies. This will help ensure that data are generated in a coordinated fashion and are focused on the primary areas of concern associated with pesticide use.

SELECTED REFERENCES

Mayer, F.L., G.F. Krause, D.R. Buckler, M.R. Ellersieck and G. Lee. 1994. Predicting chronic lethality of chemicals to fishes from acute toxicity test data: Concept and linear regression. *Environmental Toxicity and Chemistry* 13:671-678.

Science Advisory Board, USEPA. 1990a. Reducing risk: Setting priorities and strategies for environmental protection. EPA SAB-EC-90-021. Washington, DC.

Science Advisory Board, USEPA. 1990b. The report of the Ecology and Welfare Subcommittee: Relative risk reduction project, reducing risk. Appendix A. EPA SAB-EC-90-021A. Washington, DC.

Science Advisory Board, USEPA. 1990c. The report of Human Health Subcommittee: Relative risk reduction project, reducing risk. Appendix B. EPA SAB-EC-90-021B. Washington, DC.

Science Advisory Board, USEPA. 1990d. The report of the Strategic Options Subcommittee: Relative risk reduction project, reducing risk. Appendix C. EPA SAB-EC-90-021C. Washington, DC.

Sun, K., G.F. Krause, F.L. Mayer, M.R. Ellersieck and A.P. Basu. 1994. Predicting chronic toxicity based on the theory of accelerated life testing. EPA/600/R-94/058. USEPA, Gulf Breeze, FL.

Thomas, L. M. 1987. Environmental decision-making today. *Environmental Protection Agency Journal* 13:2-5.

USEPA. 1992. *Framework for Ecological Risk Assessment.* EPA/600/R-92/001. Washington, DC.

Chapter 2: RISK CHARACTERIZATION

Over the past few years there has been an emphasis within the United States Environmental Protection Agency (USEPA) and other government agencies (CEPA, 1992; National Research Council [NRC], 1992) to improve ecological risk characterization. In February 1992, USEPA's Risk Assessment Forum published a report that provides a framework for conducting and evaluating ecological risk characterizations (Figure 2-1). The process begins with a problem formulation (or scoping) phase, followed by analysis of both effects (stressor-response relationships) and exposure. In risk characterization, exposure and effects information is integrated into a risk estimate, which is provided to the risk manager along with a description of the major uncertainties and an interpretation of the ecological significance of anticipated effects.

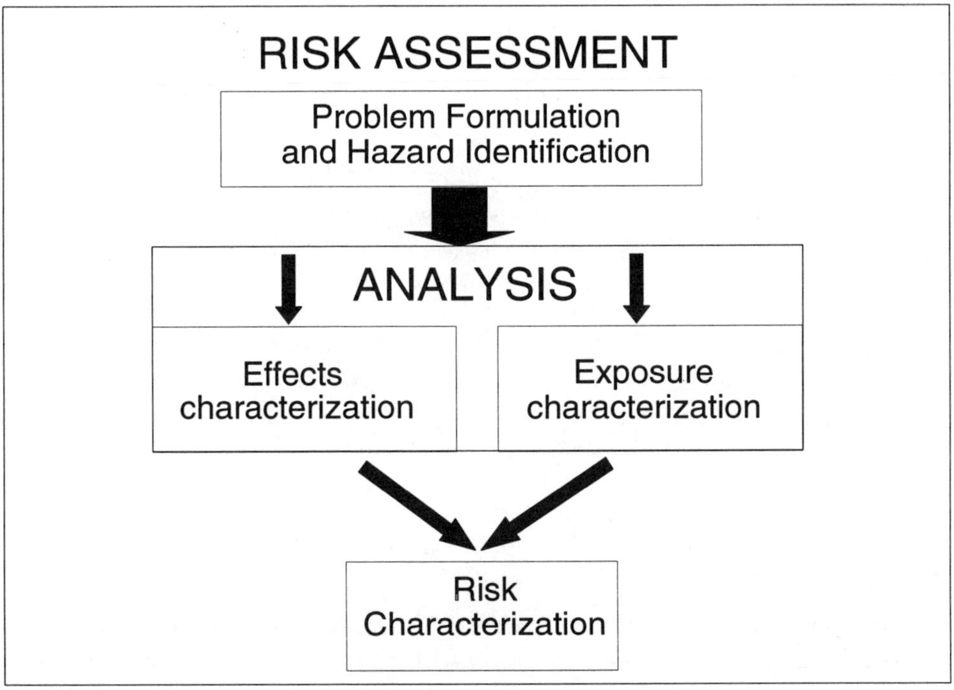

Figure 2-1. The risk assessment process as adapted from the USEPA's Risk Assessment Forum (USEPA, 1992)

In early 1992, a USEPA workgroup was convened to review the use of ecological fate and effects data in the pesticide risk assessment process in order to focus resources on the most cost-effective and value-added information necessary for regulatory decision making. The publication of the report from the Ecological Effects and Environmental Fate Task Force (Appendix A), referred to as the **new paradigm** (see Introduction), has raised a number of questions in the scientific community. The essence of the new paradigm is that initial regulatory decisions will be made earlier on the basis of estimates of exposure through the use of models, responses from single-species laboratory toxicity tests, and in some cases a limited number of field studies. The concern is that uncertainty in these regulatory decisions will be relatively great due to a de-emphasis on field studies. However, while there has been a de-emphasis on field testing, a new mitigation and monitoring phase for characterization of reductions in exposure has been added. For the most part, this paradigm has been taken as a given, and we have attempted to review and develop criteria for characterizing risk and also for validating these criteria from the existing data bases for pesticides. The Aquatic Dialogue Group fully recognizes the current state of flux in the risk assessment process. Continuous evolution is expected as new tools for adverse effect characterization and prediction of expected environmental concentrations (EECs) are developed.

The process takes place in several stages, the first of which is designed to allow a rapid characterization of those pesticides that pose little significant risk to the aquatic environment. By design, this step is conservative. The preliminary risk characterization is followed by a series of progressively more refined steps. In addition to separating pesticides that do not present a significant risk, the refined risk assessments help identify, for each pesticide of potential risk, the use patterns of greatest potential risk. Pesticides, uses, and sites/regions can then be identified for mitigation measures.

In addition, throughout the life of a pesticide's use, reevaluation of a product's potential impact will occur on a periodic basis. At that time, additional studies of exposure or effects may be desirable to refine the risk estimates. Within the new paradigm, a critical regulatory question is, "When do the refined risk characterizations trigger the implementation of mitigation?" If adverse effects such as fish kills are occurring, USEPA cannot wait for extensive refinement of the risk characterization prior to implementing mitigation or making a regulatory decision. Conversely, if risk is not clearly defined, requiring extensive mitigation prior to adequately quantifying the risk is inappropriate. This additional information should focus on increasing the accuracy and precision of the risk characterization and could include more refined exposure characterizations (*e.g.*, landscape factors) or effects studies such as additional laboratory studies or simulated field studies (*e.g.*, outdoor microcosms).

The estimation of pesticide exposure concentrations for aquatic organisms is an important part of the aquatic risk assessment process. To be consistent with the new paradigm, the risk assessment process described here emphasizes computer modeling to generate EECs rather than focuses on monitoring estimates of environmental concentrations. However, field and monitoring studies will continue to play a role in the characterization of aquatic exposures. These studies, rather than generally being required to make a registration or reregistration decision, will be used to validate computer modeling estimates, to help characterize actual pesticide concentrations in natural waters, and to determine the effectiveness of mitigation methods in reducing the exposure of aquatic organisms to pesticides.

Currently, EECs are developed in a variety of ways. Some are developed through actual measurement of water concentrations in field or monitoring studies; some are derived from "back-of-the-envelope" highest exposure scenario calculations with the assumption that various conservatively estimated percentages of the applied material go directly into a body of water; others are developed from "reasonable high exposure" or "highest exposure" computer modeling scenarios. However, these values all represent single EECs whose quantitative likelihood of occurrence is unknown because they have not been placed in a probabilistic framework that is necessary to capture the highly variable nature of actual pesticide concentrations in the environment. Procedures for estimating exposure should take the natural variation into account as much as possible by providing distributions of EECs rather than single EEC values. Such probabilistic EEC distributions over multiple years at one site, as well as over multiple sites, can be used to estimate how frequently the maximum or average concentrations of pesticides will exceed any given threshold for toxicity to aquatic organisms.

This chapter provides a discussion on the toxicity data requirements as they fit USEPA's new paradigm, followed by a discussion on the validation of the tiered probabilistic refined risk assessment process. A description of EECs that will be available for the characterization of risk to aquatic organisms is presented in Chapter 3.

2.1 PROBLEM FORMULATION AND ADVERSE EFFECT IDENTIFICATION

2.1.1 Problem Formulation

As described by USEPA's ecological risk assessment framework (USEPA, 1992), problem formulation is a planning and scoping process that establishes the goals, breadth, and focus of the risk assessment. Included in problem formulation are a description of the environmental values to be protected (the assessment endpoints), the data needed, and the analyses to be used. The goals and breadth of pesticide ecological risk assessments under the new paradigm are largely constrained by policy and resource considerations within the Office of Pesticide Programs (OPP). For example, the decision to minimize the use of field testing methods in risk assessment and the reliance on ecotoxicological single-species toxicity tests for estimating effects are, in large part, a result of policy decisions resulting from the large number of pesticides requiring registration or reregistration and the limited time and staff resources available to address these requirements. The approach for risk characterization of pesticides should allow tailoring the risk assessment for higher levels of complexity if regulators are required to evaluate more specific-use patterns or effects on specific organisms or ecosystems of concern.

2.1.2 Adverse Effect Identification

Aquatic ecosystems potentially at risk from pesticides are most frequently freshwater lakes, streams, and ponds, but may also include estuarine and marine systems depending on the use patterns for specific pesticides. For the following discussion, the major assessment endpoint in OPP aquatic ecological risk assessments may be considered to be protection of at least 90% of aquatic species from direct acute and chronic toxic effects. Measurement endpoints are primarily related to effects on mortality (acute effects) and mortality reproduction and growth (chronic effects), as determined from single-species laboratory toxicity tests.

2.1.3 Triggers Used for Chronic Tests

Currently, triggers are used to proceed from acute to chronic testing (Figure 2-2). Several criteria have been suggested as triggers for the early life-stage test in fish and for the invertebrate life-cycle test and the fish life-cycle test. These are listed below, followed by comments.

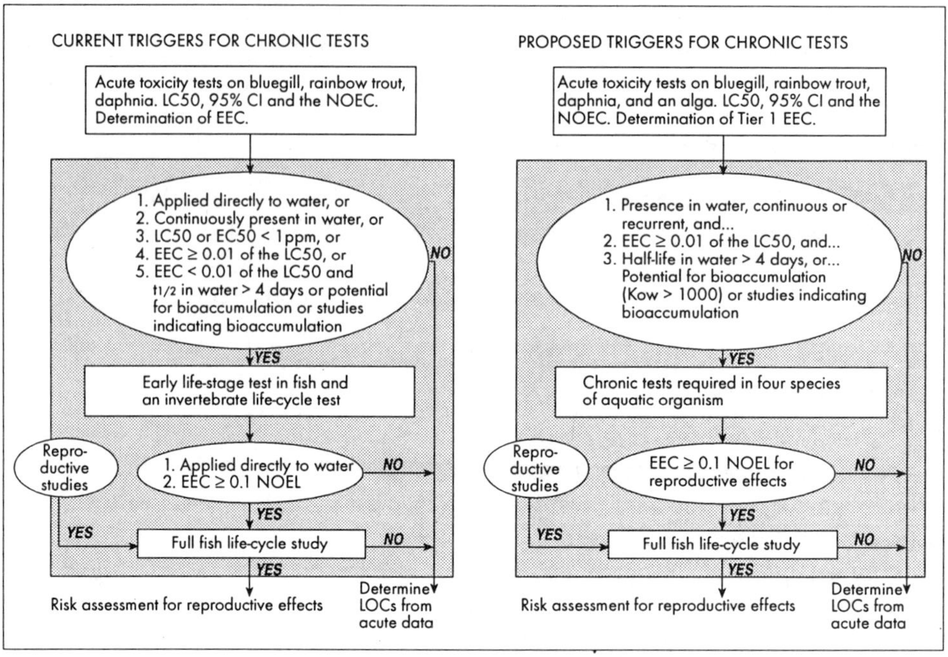

Figure 2-2. Subset of the current aquatic risk assessment decision tree showing current and proposed triggers for chronic tests (USEPA, 1984)

1. "The product is intended for direct application to aquatic environments, or" ...

   This is an appropriate trigger for chronic studies only if the pesticide has a long half-life or is used frequently. Direct application to an aquatic environment may inevitably result in high EECs that could trigger levels of concern (LOCs) for refined risk characterization.

2. ..."the pesticide is intended for use such that its presence in water is likely to be continuous or recurrent, or" ...

   This is an appropriate trigger for chronic studies.

3. ..."any LC50 or EC50 determined in acute toxicity testing is less than 1 mg/L, or" ...

   The rationale for this criterion is based on conventional toxicity ranking of chemicals. It appears not to be related to toxicity and was probably not originally derived from any scientific principles. It is recommended that this trigger be dropped.

4. ..."the EEC in water is equal to or greater than 0.01 of an LC50 or EC50 determined in acute toxicity testing, or" ...

   This criterion is designed to protect against the likelihood of chronic toxicity on the basis of acute toxicity data. The use of a 100 x factor is supported by published analyses on acute-to-chronic ratios from aquatic toxicity studies (Sloof *et al.*, 1986; Giesy and Graney, 1989). *However, it is felt that this may result in unnecessary testing of pesticides that are not continuously present, persistent, or bioaccumulative (triggers 1, 2 and 5). Therefore, it is recommended that this trigger be linked to repeated application, continuous presence, persistence, or the potential for bioaccumulation.*

5. ..."the EEC is less than or equal to 0.01 of the LC50 and the half-life in water is greater than 4 days or the physicochemical properties indicate cumulative effects, or" ...

   The use of a safety factor greater than 100-fold for aquatic risk assessment is not supported in the literature and may be unnecessary to trigger pesticides into chronic testing. This half-life should be conservative, i.e., the longest abiotic hydrolysis $t^1/_2$ *in the pH range of 5 to 9. The assessment of half-life is appropriate, but the use of 4 days as a trigger should be assessed in relation to the use-pattern. Potential for bioaccumulation is assessed via the* K_{OW} *(>1000). This is the same trigger used to initiate a fish bioaccumulation study and is appropriate as a trigger for a chronic study.*

6. ..."the fish life-cycle study is triggered by the product being directly applied to water and if the EEC ≥ 0.1 of the NOEL [no observed effect level] from the fish early life-stage or invertebrate life-cycle test" (USEPA, 1984).

   The use of effects in an early life-stage test on juvenile fish as a trigger for a full life-cycle test in fish would be justified only where additional evidence such as structure-activity relationships, results from mammalian or avian studies, or the invertebrate life-cycle test indicated reproductive effects.

The following (Figure 2-2) are recommended as triggers for fish early life-stage or invertebrate chronic studies:

- The pesticide is intended for use such that its presence in water is likely to be continuous or recurrent,
- the EEC ≥ 0.01 of the LC50 or EC50, and
- the half-life in water is greater than 4 days, or the physicochemical properties indicate cumulative effects (e.g., K_{OW} > 1000).

Other tests such as the life-cycle studies on fish should be triggered by indications of reproductive effects, from structure-activity characterizations or the results of avian or mammalian tests carried out in response to other sections of the risk assessment process. The difficulty of conducting the fish life-cycle tests also suggests that a more appropriate test of fish reproductive effects be developed.

2.2 EFFECTS CHARACTERIZATION

The current aquatic testing requirements for registration consist of acute studies on three organisms (two fishes and one daphnid) and, if triggered, an early life-stage test with one fish and a full life-cycle daphnid study. In order to implement a probabilistic approach for risk characterization, additional acute tests would be required to better describe the distribution of susceptibility in the "universe" of species. This additional testing may be accessible through harmonization with other testing requirements such as those being proposed by the Great Lakes Water Quality Guidelines (GLWQG, 1993), which are generally consistent with the data requirements for USEPA water quality criteria (Stephan *et al.*, 1985). The GLWQGs recommend that acute tests are required on eight species:

- Vertebrates—one species of salmonid fish, one species of warm water fish species (*e.g.*, bluegill), and either another fish or an amphibian species (three species total)
- Invertebrates—one planktonic crustacean (*e.g.*, Daphnia), one benthic crustacean (*e.g.*, crayfish), one insect (*e.g.*, mayfly), one invertebrate from a phylum other than the above (*e.g.*, rotifer, annelid worm, or mollusc), and one order of insect or one phylum not already tested (five species total)

In addition, the GLWQGs recommend that three chronic tests be conducted on three families of animals (one fish, one invertebrate, and at least one species of acutely sensitive freshwater animal). Okkerman *et al.* (1993) recommend that four chronic tests be required. The guidelines also suggest that a test on a freshwater alga or vascular plant is desirable. The suitability of an alga or *Lemna* sp. as a surrogate for all aquatic macrophytes is questionable, and the need for additional plant data and the usefulness of this data must be assessed, particularly in the case of herbicides. Species for testing have been reviewed and assessed by Blanck *et al.* (1984), Doherty (1983), Kenaga (1978), Kimerle *et al.* (1985), Mayer and Ellersieck (1986), and Sloof *et al.* (1983).

2.2.1 Data Requirements

For a probabilistic approach to effects characterization, it is recommended that there be eight acute studies. The species tested should include at least two species of fish, one invertebrate, and one aquatic plant (macrophyte or alga). The other four species should be selected from the group(s) of organisms observed to be the most sensitive after the initial tests are complete. Tests on marine species can be used with data from freshwater organisms. If chronic studies are required, then it is important to understand the nature of these chronic tests. The purpose is to provide a characterization of the effects of chronic exposure over a relevant period of the lifespan of the organism, whichever is shorter. These tests do not necessarily require the use of early life stages, nor should they necessarily assess endpoints other than growth. In choosing appropriate species for the chronic tests, it is recommended that one fish, one invertebrate, and one plant (macrophyte or alga) be tested. Also, it is recommended that one additional organism be chosen using the most sensitive group as identified in the suite of acute toxicity tests.

While a core data base is necessary in order to compare relative toxicities of pesticides at an initial screening level, it is appropriate at higher levels of effects characterization to consider the actual use pattern and physical and chemical parameters of the pesticide in relation to data generation. Exposure is a function of how much/how often the pesticide is present in the environment, how much is available to the organism, and to what extent nontarget organisms will come in contact with the pesticide. The last factor depends on the life history, movement, and demographics of the organism. The first two factors relate to such parameters as the use rate and application method, physical and chemical properties of the pesticide, and runoff and drift rates. The goal of customizing tests to fit actual use patterns would be to design, perform, and present studies that enable both the risk assessor and the risk manager to make more informed decisions. While it may be most desirable to perform additional customized testing, additional useful data can be obtained from standard acute studies if the data are needed.

In some cases, it is known that reciprocity between toxicity and cumulative exposure from pulsed exposures is not linear for most pesticides. For instance, pulsed exposures to greater concentrations may be less toxic than longer term continuous exposures to lesser concentrations. Although these exposures result in the same integrated concentration-time exposure (ppm x min), they may not account for pharmacokinetic or repair processes which control the effects that are observed. Currently, there is little information available to allow these types of analyses to be conducted. For some pesticides, such as organophosphorous insecticides, the acute-to-chronic ratio is essentially equal to one, and the sum of the product of duration and intensity of exposure does display reciprocity. Therefore, an analysis of the type proposed could be made to work for these types of pesticides with data that are currently available. However, some of the organophosphorous pesticides cause effects through metabolites that result in delayed toxicity which does not display reciprocity. For these pesticides or those for which these types or mechanisms are not understood, special short-term exposure studies with appropriate follow-up may have to be conducted.

An example of customizing a standard acute test would be to add additional exposure concentrations, time periods, or endpoints for characterization. If the pesticide in question is used once per year and has a half-life in water of 1 hour (h), assessing mortality on a shorter time scale (*e.g.,* 4 h, 8 h, etc.) in a 48- or 96-h study would allow a probability distribution to be created which compares short-term exposure to short-term toxicity. This consideration of the temporal pattern of mortality as it relates to EEC calculations provides the risk manager with a more complete picture of the risks associated with the actual use of the pesticide.

Timing of use can also have an impact on the type of studies that should be performed. If a pesticide is used only once or very infrequently during the year and is not persistent, then the pesticide would not be likely to have chronic effects. Given this scenario, it may be better to focus resources on additional acute studies with more species than on chronic studies.

In order to provide more insight into ecosystems at risk and thus to consider testing scenarios that are suitable for particular areas of concern, effects characterizations should include information on the regional distribution, timing of use, and extent of use of the pesticide. Consideration of the regional distribution of use enables the registrant to test the most appropriate species. For example, if a pesticide is used only in the extreme southern part of the U.S., then species other than rainbow trout (*e.g.*, catfish, darters, sunfish) may be more appropriate to test, in addition to the bluegill. Similarly, if the pesticide is used during a time of the year in an area that experiences migration of species (*e.g.*, salmon), it could be important to determine the impact of the use of the pesticide on that species. Geographical Information Systems (GIS) techniques could be employed to more fully understand the actual use of a pesticide and its relation to valued ecological resources as endangered species.

2.2.2 Supplementary Effects Information

In order to get the best data and gain the most useful information for the risk assessor and the risk manager, it is important to be able to focus resources where they are most needed. In addition to the studies needed to determine the probability distribution of effects for the refined risk characterization, more focused studies could be performed in order to better define the toxicity profile in the environment. While these proactive tests would not serve to replace the required guideline studies, they should be considered during the risk characterization and risk management processes in order to better understand the actual risk to the aquatic environment. An example of how additional information could be beneficial relates to use pattern. If the pesticide is used once a week for a number of weeks then, in addition to the standard acute toxicity test, it would be appropriate to utilize a pulse-dosed system to mimic the incremental inputs of the pesticide over time.

Another example relates to the pesticide to be tested. Typically, the toxicity value used in effects characterization is generally that of the technical material. Current requirements for performing laboratory studies with the formulated end-use product are based on the toxicity versus the expected environmental concentration (*e.g.*, EEC $\geq$ 0.5 LC50 or EC50 of technical material), the intended use (*e.g.*, labeled for use directly into aquatic environments), or the presence of ingredients in the formulation that would enhance the toxicity. When toxicity information on the end-use product is available, it should be considered in the refined risk characterization along with the technical material. Ingredients in formulations can change the availability of the active ingredient (*e.g.*, making it more or less soluble) or may themselves have an effect on nontarget organisms. In addition, some products are formulated in such a way as to limit the release of, and thus exposure to, the active ingredient. For example, there are time-release aquatic herbicides that are formulated to sink to the bottom of a water body and slowly release the active ingredient into the root zone. This results in much less of the active ingredient being available in the water at any one time than would be assumed by the use rate, thus limiting acute risk to nontarget organisms. However, such a release pattern may increase continuous exposure and create a chronic scenario. While relying on the information concerning a formulated product is more realistic and may seem like the better alternative, the difficulty is that as ingredients in a formulation change, new studies would and should be required. Continual retesting could become cost prohibitive and could slow research and development efforts to improve formulations.

Another consideration relating to physical and chemical parameters is when the pesticide binds to soil and sediment. Laboratory tests carried out in "clean water" are usually conservative or "highest exposure scenario" to nontarget organisms such as fish and daphnids due to the lack of natural partitioning to sediment and the constant availability of the pesticide. While assumptions are often made that once a pesticide enters a water body it is taken up by aquatic organisms, parameters such as bioavailability can and do play a crucial role in the actual exposure scenario. The bioavailability of pesticides from both suspended and bottom sediments in combination with toxicological effects data can provide information about the risk associated with chemical/sediment interactions. However, there are currently no requirements to perform studies with sediment-dwelling or benthic organisms. These nontarget species could be at risk from a pesticide that binds to sediments, especially if it is not degraded quickly. While the historical data base for sediment tests is not as extensive as that for fish such as rainbow trout and bluegill, there is a need to consider the impacts on sediment-dwelling species, and further work should be focused on developing a data base and acceptable protocols for performing sediment studies and assessing and interpreting the data.

While the additional, non-guideline testing discussed above is not required, it is recommended that additional species (when available) testing and focused studies such as formulated product testing, pulse-dosed exposures, and other studies that help define the toxicity profile of a pesticide as it relates to use pattern or physical-chemical parameters should be utilized by the risk managers to aid in the understanding of and risk management decisions concerning a pesticide.

2.2.3 Requiring Effects Data after Mitigation

Requiring additional effects data after mitigation has been implemented must focus on characterizing the residual risk. The following types of questions should be addressed prior to any additional data requirement:

- What is the specific objective of the study being required?
- How will the data from the study be analyzed and presented?
- Are the procedures in place for conducting such a study, or is method development work required?
- What is the "value-added" associated with conducting the study?

The specific types of data that may be required should be evaluated on a case-by-case basis. Additional laboratory studies (such as sediment toxicity, pulsed-dose exposures, and additional species testing) should not be required unless those studies can provide predictive information directly applicable to risk characterization and will be used by risk managers in regulatory decisions. Similarly, the objective of comparative or simulated field studies should be to address specific questions useful in characterizing the residual risk. Simple simulated field studies using microcosm-size systems may address specific questions which can greatly enhance the predictive nature of the risk characterization. The value of these studies will be greater if they are not designed to refute the presumption of risk, but rather to identify significant effect and no-effect levels and to establish time-concentration-response relationships for important components of the ecosystem. Workshops have been held which outline the methods, approaches, and flexibility of such systems (SETAC-Europe, 1991; Crossland *et al.*, 1992; SETAC, 1994). Various techniques, such as *in-situ* toxicity tests and bioassays with components/matrices removed from microcosms, can also be used to answer specific questions and test hypotheses.

2.2.4 Current Levels of Concern (LOCs)

Currently, risk assessment decisions involving aquatic organisms are based on the ratio of a single EEC to a single reference effect level of the pesticide. This relationship is known as the risk quotient, Q. The maximum acceptable risk quotient (such as 0.1, 0.5, 1) is the level of concern (LOC). At present, for acute aquatic effects, no regulatory action is necessary if the EEC < 0.1 LC50 for the most sensitive species tested (Figure 2-3). A pesticide falling in this category would be considered to have a low risk to aquatic organisms. However, if 0.1 LC50 ≤ EEC < 0.5 LC50, the pesticide would be considered for a restricted-use status. This means that only a certified pesticide applicator would be able to apply the pesticide. This restriction is intended to limit the distribution and use of the pesticide to some extent and to ensure compliance with label instructions, since use by untrained

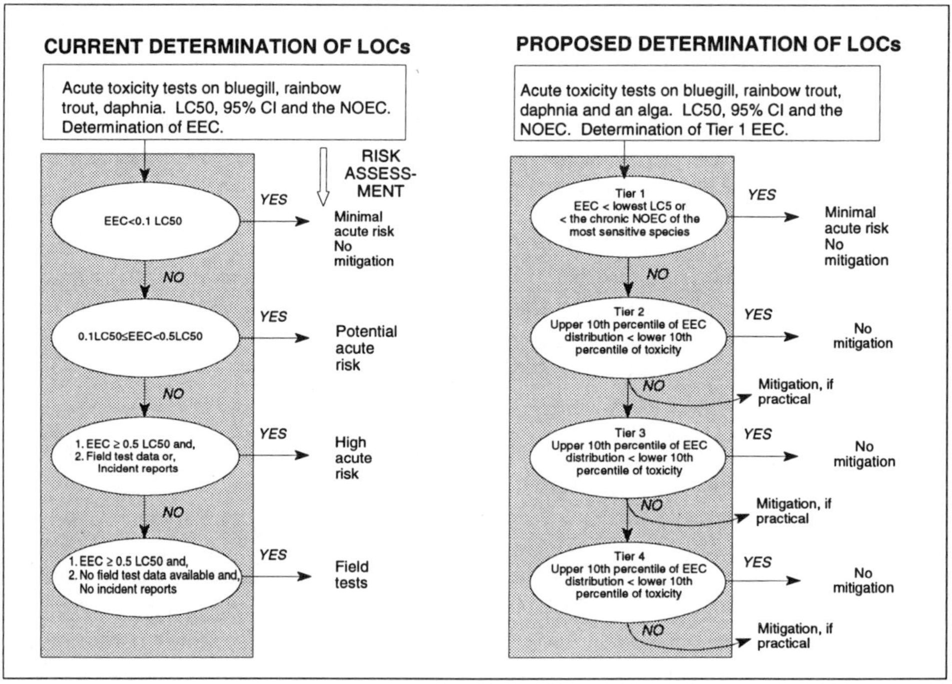

Figure 2-3. Subset of the proposed new aquatic risk assessment decision tree showing the current and proposed LOCs

individuals would likely be prohibited. The risk to aquatic life from the use of a pesticide is deemed to be high when the EEC $\geq$ 0.5 LC50. A pesticide with this level of concern would not only be classified as restricted use, but would also be considered for further regulatory action, such as mitigation, special review, or cancellation.

An assessment of field incident data and results of field studies could be used to judge the appropriateness of the reference effect level and the LOC values currently used (Figure 2-3). A preliminary analysis was performed (Section 2.4) for this discussion, but the paucity of data from field studies and the lack of consistency in the sampling of residues from field incidents did not provide meaningful conclusions.

While these reference effect levels (such as the LC50s) and LOCs have been accepted and used historically, there is concern that they may be either over- or underprotective depending on the pesticide. Using a fixed fraction (0.1 or 0.5) of a point estimate of the LC50 to determine an LOC does not consider possible variations in the slope of the concentration-response line between pesticides. In the example in Figure 2-4, both concentration-response lines have the same LC50 but different slopes. Line 1 has a slope of 1 (probit units per log cycle), and at a concentration of 0.1 of the LC50, a mortality of about 16% would be expected. This could be even higher if the upper 95% confidence interval of this estimate were taken into account. For environmentally or economically important species, mortality at this level could be considered unacceptable. Line 2 has a steeper slope, and the use of 0.1 of the LC50 would result in a mortality lower than 1%, and thus may give adequate protection. Use of a fraction of the LC50 on pesticides with a slope steeper than this may be unnecessarily overprotective. On this basis, it would appear that the use of a point estimate of a low-response concentration calculated from the slope of the line (for example, LC5 or some other level of response) would be more appropriate. In choosing a level of response, consideration must be given to balancing the need to choose a low level of effect (unlikely to affect the population) with the uncertainties associated with excessive extrapolation of the concentration line. Selection of an appropriate level of protection is best carried out by a retrospective assessment of the relationship between LC50 and the slope of the concentration-response line.

2.2.5 Use of Existing Data to Validate Current LOCs: Slope of the Concentration-Response Line

The criterion of 0.1 LC50 is the current regulatory threshold. This threshold is based on the presumption that significant acute toxicity will not occur at this level of exposure. This level of exposure will depend on the slope of the concentration-response curve. USEPA is compiling a data base of toxicity values submitted by registrants of pesticides (Urban, 1993). The data base currently contains about 4,400 records, 335 of which have slope values entered and were based on studies in aquatic organisms. These data have been analyzed to assess the distribution of the slopes and the validity of the use of the 0.1 LC50 as a measure to assess the EEC.

Examination of the relationship between log 0.1 LC50 and other log LC values such as the log LC5 (Figure 2-5) showed that, for the most part, a linear relationship existed between these values. However, the data in the graph are potentially misleading because of the use of the log scale (to allow display of all the data points). The range of the data spanned 14 orders of magnitude and the confidence intervals (Figure 2-5), although appearing narrow on the graph, spanned a considerable range (approximately 100x). There were also several data points that fell well outside the general grouping. Thus, a fraction of the LC50s will be a poor estimator of low-concentration effects for some pesticides. From these observations, it is clear that an LOC criterion that can take these exceptional relationships into account must be established.

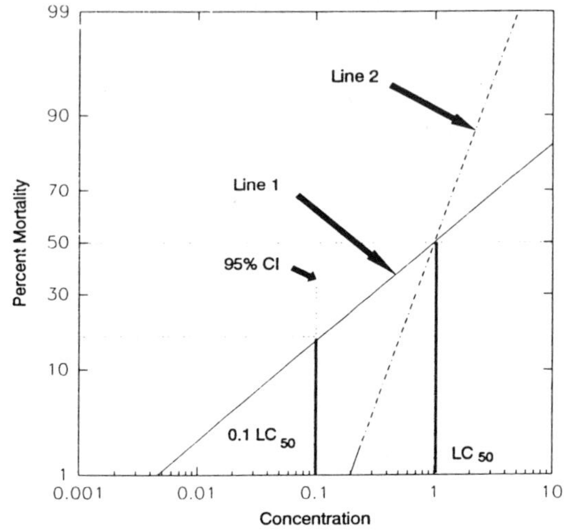

Figure 2-4. Comparison of two concentration-response curves with the same LC50 but different slopes

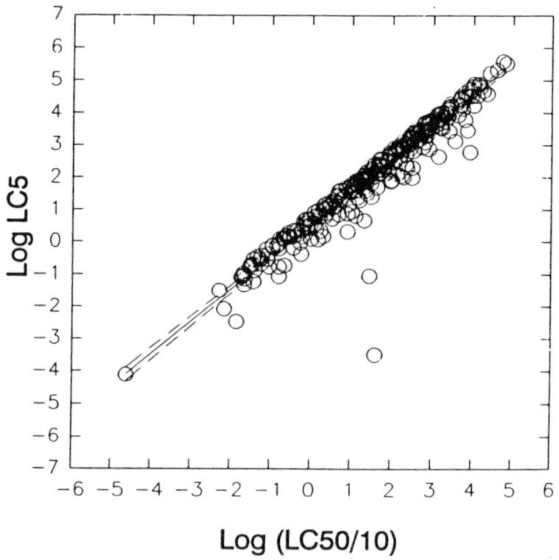

Figure 2-5. The relationship between 0.1 LC50 and LC5 for 335 pesticides tested in aquatic organisms (95% confidence intervals shown by dashed lines)

Another way of assessing the data was via a comparison of the LC5 (as calculated from the slope of the concentration-response line) and 0.1 LC50. A similar approach has been used in the development of water quality standards by USEPA (Delon *et al.*, 1991). The LC5 was chosen as a reference point because it is the concentration below which one would not expect to be able to discern any acute toxic effects in the organisms exposed under field conditions. Thus, assuming field exposure, if 0.1 LC50 is greater than the LC5 (*i.e.*, the ratio of [0.1 LC50]/[LC5] > 1), one could expect observable acute toxicity to occur. These data (Figure 2-6) show that, for the majority of substances (90%), the ratio is less than one, meaning that 0.1 LC50 is protective or overprotective of the LC5 in these cases. On the other hand, a significant number of pesticides (10%) showed ratios that were greater than 1. In these cases, 0.1 LC50 may be underprotective of the LC5 in these organisms.

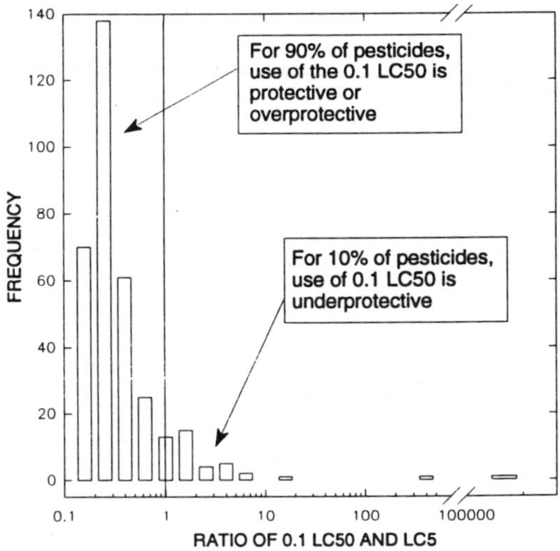

Figure 2-6. Frequency distribution of the ratio of the 0.1 LC50 and the LC5 for 335 compounds tested in aquatic organisms (data from Urban, 1993)

A more detailed analysis of these data for herbicides and insecticides in fish and crustaceans (Figure 2-7) showed that the slopes of the concentration-response lines were distributed in a log-normal manner. This analysis also showed that, for the two groups of pesticides tested in these organisms, the slopes of the concentration-response lines were all greater than 1. Table 1 shows the arithmetic and geometric means calculated from these distributions. The lower 95th percentiles (geometric transformation) for the slopes were, in all cases, greater than 1. Thus, for this subset of the data, *if slope data had not been available*, an assumed worst-case slope of 1 would have been protective of aquatic organisms in 95% of cases. If the slope is ≤ 1, then 0.1 of the LC50 corresponds to a mortality of $\geq 16\%$ (Figure 2-4).

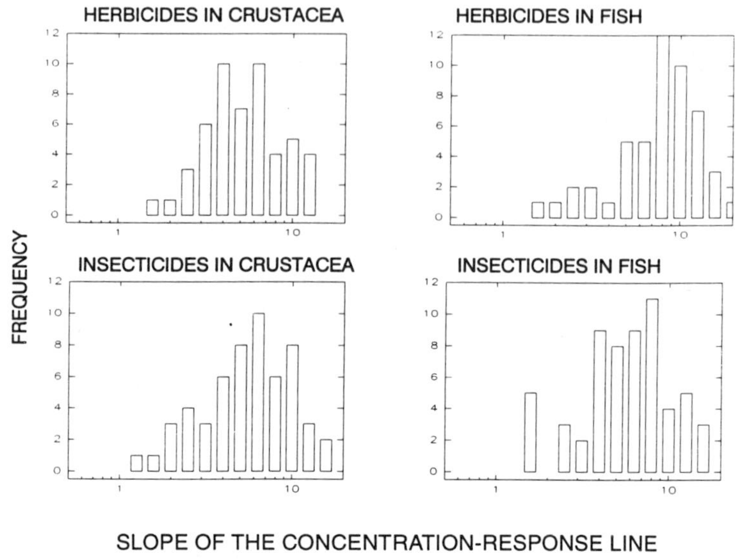

Figure 2-7. Frequency distribution of slopes of concentration-response lines for herbicides and insecticides in fish and crustaceans

TABLE 2-1. Geometric Analysis of Distributions of Slopes of Concentration-Response Lines for Insecticides and Herbicides in Crustaceans and Fish

Combination	Mean Slopes	Upper 95th	Lower 95th
Insecticides in fish	5.0	8.6	1.4
Herbicides in fish	6.7	10.1	3.3
Herbicides in crustacea	4.7	7.9	1.5
Insecticides in crustacea	4.8	8.4	1.3

The distribution of the slopes (geometric transform, all pesticides and organisms) for all the data (Table 2-2) suggested that an assumed (worst-case) slope of 0.85 could be used to calculate a low LC value, such as an LC5 that would be protective of all test organisms from all pesticides in 95% of the cases. In both these cases, it is unlikely that this type of worst-case assumption will be necessary. It is unlikely that a registrant will not have good acute toxicity data from which slopes can be calculated.

TABLE 2-2. Slopes of the Concentration-Response Lines for All Species and Pesticides (N=335)

Geometric mean of slopes	4.67
Geometric SD of slopes	1.95
Upper geometric 95th percentile of slopes	8.50
Lower geometric 95th percentile of slopes	0.85

Analysis of the above data suggests that the triggering of a preliminary LOC when the EEC > 0.1 LC50 would not be conservative enough in 10% of cases. A retrospective assessment of current data (Section 2.4.4) suggests that a criterion based on the LC5 would capture some 75% of pesticides for Tier 2 characterization. Use of a more conservative (lower) LC value may force extrapolation of the concentration-response line to regions of nonlinearity. It is therefore recommended that the slope data be used to calculate an estimate of a low level of effect (LC5) and that this be used for comparison to the EEC. These data are generally already available and have been partly compiled in a data base by the USEPA. Further, it is recommended in the general sense that LC values always be published in reports or in the literature with confidence intervals and the slope of the concentration-response line.

2.2.6 Use of Probabilistic Endpoints for Effects Characterization

Expressing the results of a refined risk characterization analysis as a distribution of toxicity values rather than a single point estimate is an approach presently being used by the Dutch government (Health Council of the Netherlands, 1993). A major advantage of this approach is that it uses all relevant single-species toxicity data and, when combined with exposure distributions, allows quantitative estimations of risks to aquatic organisms. An illustration of the principle is shown in Figure 2-8 C, adapted from Cardwell *et al.* (1993). This approach is based on two assumptions:

1. that the distribution of acute susceptibility represents the "universe" of species and is log-normal with respect to concentration (shown as a cumulative frequency distribution in Figure 2-8 A) and
2. that the chronic toxicity of a pesticide is log-normally distributed (Figure 2-8 B).

The degree of overlap of the exposure curve with the effects curve can be used to estimate the probability that a certain percentage of aquatic species may be adversely affected. There are several outcomes of this approach. The distribution of EEC-1 (Figure 2-8 C) has little overlap with the chronic toxicity distribution, suggesting that there will be little impact from this situation. The distribution of EEC-2 is such that more overlap is present.

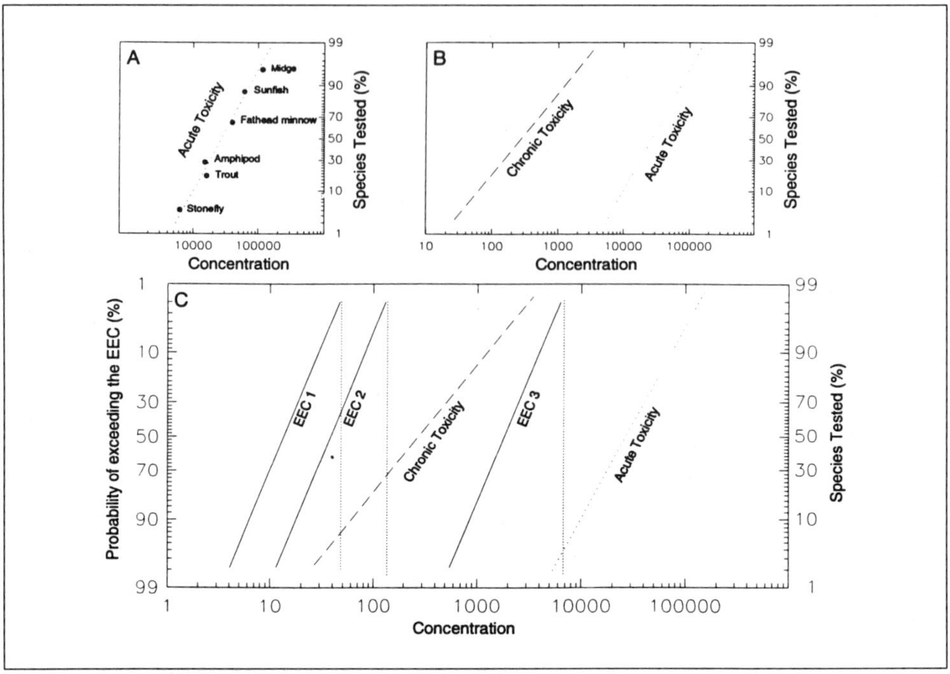

Figure 2-8. Comparison of EECs for a substance with concentrations causing acute and chronic toxicity

In this situation, it is likely that chronic effects will be observed unless mitigation measures can move the EEC distribution to the left or increase the steepness of the slope. This could be done by identifying those environmental situations that lead to the high EECs or by identifying those very sensitive species and restricting the use of the product in these environments or in areas where the sensitive species occur. In the case of EEC-3 (Figure 2-8 C), chronic effects will likely be observed in almost all instances, and major mitigatory steps may be required. In the Dutch system, distributions of chronic toxic effects (no observed effect concentrations [NOECs]) are used to estimate the concentration of a chemical at which only 5% of the species would be affected (Health Council of the Netherlands, 1993). In other words, at the chosen level of protection, there is a chance that 5% of the aquatic species may be affected. In using overlaps of distributions, there is a tacit assumption that protecting a certain percentage of species for a certain proportion of occasions will also preserve ecosystem structure and function. The other assumption is that the selection of species used in the tests is a good surrogate for the "universe" of species in the environment.

There are some limitations to this approach (Health Council of the Netherlands, 1993). For example, the choice of protection level (*e.g.*, 90% of species) may not be socially acceptable. Some may view 90% as being overprotective, while others may find that level of risk unacceptable, especially if the 10% of potentially affected species includes organisms of high ecological, commercial, endangered species, or recreational significance. In addition, risks of persistent, bioaccumulative chemicals to species at the top of the food chain may not be sufficiently considered by this approach alone. In the former situation, these species could be identified and appropriate mitigation measures taken.

Another point requiring further evaluation is the number of data points (test results) required to define the species distributions for either acute or chronic effects. Additional acute tests beyond those presently required may be needed. Resources needed for additional chronic tests could be obtained by deleting the requirement for a full fish life-cycle chronic test with one species, and instead conducting early life-stage-type tests with several species.

2.2.7 Dealing with Uncertainty in the Susceptibility (Effects) Distribution

Use of a susceptibility distribution in developing criteria requires that the species selected for these tests be truly representative of the "universe" of species that may be present in the environment. Essentially, this means that in addition to one very sensitive species, a relatively insensitive species should also be included in the range of species tested. The former is required to make sure that species representative of high sensitivity are present (thus avoiding Type I errors), while the latter is required to ensure that the distribution is not artificially skewed to overrepresent the highly sensitive species (thus avoiding a Type II error). The appropriateness of the distribution is obviously difficult to confirm experimentally, although the concept is similar to the cluster approach recommended by Mount (1982) and Birge and Black (1982). For this reason the Aquatic Dialogue Group recommends choosing species at the sensitive region of the distribution. This has the potential to skew the distribution, but if this is the case, it will be on the side of conservatism.

It is likely that organisms will respond differently to different chemicals and that responses will vary in their range. As with the EEC, that range will be reflected in the slope of the distribution line and in the degree of overlap (Figure 2-9).

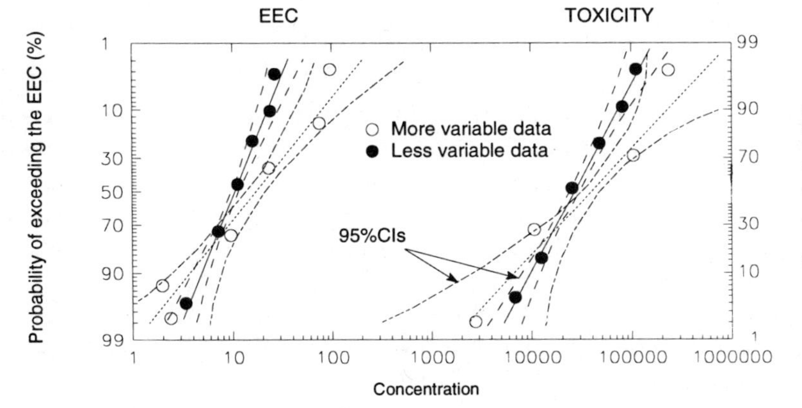

Figure 2-9. Illustration showing how uncertainty in the EEC, chronic toxicity, and acute toxicity distributions will affect the slope and overlap of the lines

2.2.8 A Tiered Approach to Risk Assessment and Need Identification for Mitigation

The process of risk assessment should be tiered to reduce the need for unnecessary assessments and delays in the regulatory process. The outline of this approach is explained in more detail in the following sections. The approach is to begin by using a conservative criterion for judgment. As the assessment is refined with more likely estimates of toxicity, exposure, and EECs, less conservative and more realistic criteria can be used (Figure 2-10).

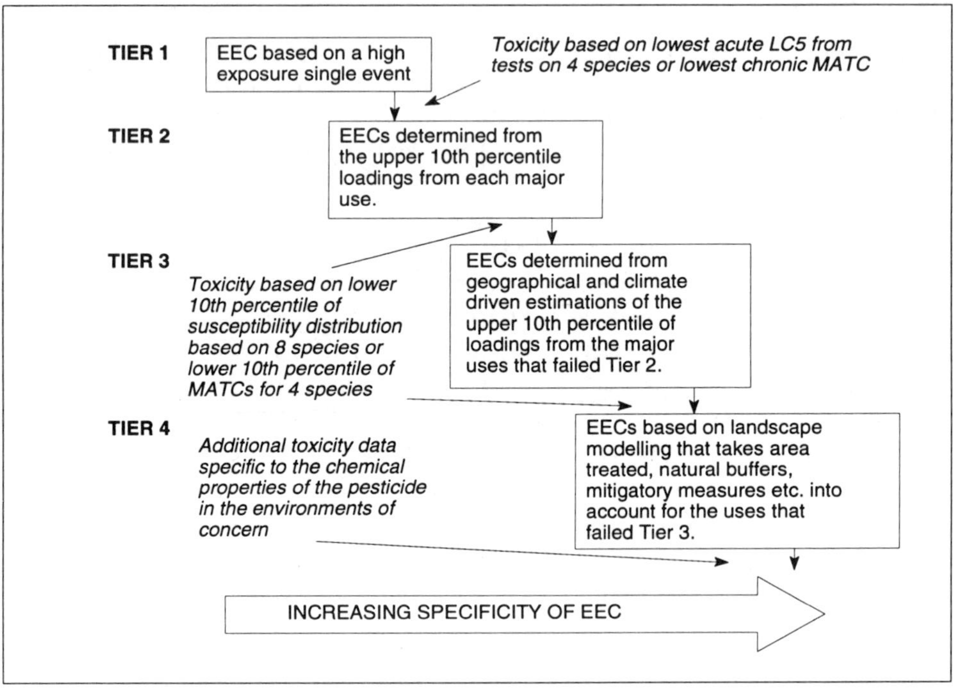

Figure 2-10. The proposed tiered approach to assessing risks from pesticides

The tiered approach makes use of the LOCs illustrated in Figure 2-3. Toxicity data that are currently available for Tier 1 would be as follows: Acute concentration-responses in three freshwater aquatic animals and one plant (and in some cases, saltwater organisms), half-life in water, K_{ow}, and in some cases, chronic toxicity concentration-responses in two species of aquatic organisms. The recommended data requirements for probabilistic risk assessment would make available data from eight acute tests and, where necessary, four chronic tests. In selecting a toxicity criterion, it is recommended that the LC5 be used as the acute toxicity criterion (Section 2.3). It is recommended that the maximum acceptable toxicant concentration (MATC), NOEC, or lowest observed effect concentration (LOEC)—depending on an assessment of the studies—be used as the chronic toxicity criterion.

2.3 RISK ASSESSMENT: EEC MODELING AND TOXICITY CRITERIA FOR THE TIERED RISK ASSESSMENT

The framework for risk assessment is based on refined aquatic risk characterization methodology that utilizes modeling and takes into account the fact that computer modeling can accurately simulate the effects of some, but not all, practical mitigation methods. At present, most models focus on the "edge of the field" and are unable to take larger geographic regions into account. As discussed in Chapter 4, basin or watershed responses and the effects of mitigation strategies may have to be determined with properly designed field and/or monitoring studies (spray drift, small-scale runoff, field-scale runoff, field-scale surface water monitoring, basin-scale surface water monitoring).

EEC characterization via computer modeling is described only briefly within this framework, but more extensive treatment is presented in Chapter 3.

Tier 1: Toxicity. Acute toxicity and slope data for three species of freshwater animals (and saltwater aquatic animal, if the use pattern suggests contamination of estuarine or marine areas) are required. In addition, one toxicity test for at least one species of aquatic plant, and if triggered by persistence, K_{OW}, etc., chronic studies on four species of aquatic organisms are recommended (Section 2.2.1). The LC5 for each of the acute tests would be calculated from the slope of the concentration-response line and the lowest number used for characterization.

Tier 1: EEC. The EEC would generally be determined for a single runoff event on a high runoff site/scenario (based on direct overspray of a pond 2 m deep or 1 to 10% runoff [depending on K_{OC}] from a 10 ha field into a 1 ha pond). If the pesticide is directly applied to water, the concentration resulting from a 100% loading would be used as the EEC.

Tier 1: Assessment. This is done by simple comparison: Is the EEC < the LC5 for the most sensitive species? And, where chronic toxicity data are available or required, is the EEC < the chronic NOECs for the most sensitive species? The most conservative of the above characterizations is used. Tier 1 is used to initially screen pesticides to determine whether additional risk characterization is needed.

Tier 2: Toxicity. The recommended toxicity characterizations are as follows: Additional acute toxicity tests will be required at this point. A minimum of four additional species should be selected from the group(s) of organisms (fish, algae, etc.) observed to be most sensitive in Tier 1 data. This practice is necessary to obtain a better estimate of the distribution of susceptibility at the lower end of the distribution.

These toxicity data can then be used to estimate distributions of the LC5s. Ideally, this estimation should be done by plotting the distribution of aquatic toxicity values as a cumulative frequency exceedance curve. Each data point on an aquatic toxicity value cumulative frequency exceedance curve will have coordinates

y_i = aquatic toxicity value for species i and

x_i = fraction or percentage of species for which the aquatic toxicity value $\geq y_i$.

In practice, occasions where this amount of data is available will be rare, and it is recommended that these be plotted in rank order (as a percentage calculated from i/(j + l)) against the log of the LC5 value to obtain an estimate of the distribution.

Tier 2: EEC. The EECs should be modeled from a small number of standardized reasonable high-exposure soil/climate scenarios, each approximately representing the upper 10th percentile of runoff yield (sediment and water) for each major use. Management practices should be assumed to be poor. The average K_{OC} for the range of soil types in this scenario should be used. If a single soil $t^1/_2$ is available, it should be increased by a factor of up to 5 to account for uncertainty in the single-valued number; if average soil $t^1/_2$ is available, the upper 90% confidence limit should be calculated from the expression

$$\text{Mean} + t(Sd/N^{0.5})$$

where

Mean	= mean of the half-life values,
t	= Student's one-tailed t at 90% for the degrees of freedom,
Sd	= standard deviation of the mean assuming normal distribution, and
$N^{0.5}$	= square root of the number of observations in the mean.

These should be modeled over ≥ 36 years to give, for each specified duration of toxicity test (*e.g.*, 4, 12, 24, 48, 96 h for acute responses, 21 d for chronic responses, etc.), a distribution of estimated annual maximum average dissolved EECs for each site/scenario modeled.

For each specified duration (48 h, 96 h, 21 d, 60 d, etc.) the distribution of annual maximum average dissolved EECs should be plotted as a cumulative frequency exceedance curve. Each data point on any given duration EEC cumulative frequency exceedance curve will have coordinates

y_i = annual maximum (specified duration such as 48 h, 96 h, 21 d, 60 , 90 d, etc.) average dissolved EEC for year i at the best judgment reasonable worst-case site modeled and

x_i = fraction or percentage of years at the site in which the annual maximum average dissolved EEC $\geq y_i$.

The upper 10th percentile of the annual maximum (specified duration) average dissolved EEC should then be determined from the EEC cumulative frequency exceedance curves. The upper 10th percentile annual maximum average dissolved EEC for yearly distributions over a single site corresponds to a value that would be expected to be equalled or exceeded at the site only approximately once every 10 years over a long time (a one-in-10-year return frequency). Another interpretation is that, for any specified duration, the upper 10th percentile annual maximum average dissolved EEC corresponds to a value that, in any given future year at the site, has an approximately 10% chance of being equalled or exceeded in a period of ten years.

Tier 2: Assessment. Tier 2 is conducted to identify uses of potential substantial risk (*e.g.*, corn, cotton, turf, etc.) for each pesticide identified as of potential substantial risk in Tier 1. This is achieved by a more complex comparison of the point estimates of the upper 10% annual EEC for a high-runoff site/scenario representing a given use with the lower 10th percentile of the toxicity distribution. This can be represented in a series of questions.

- For each high-runoff site/scenario solely representing a given use, is the upper 10th percentile of the EECs (48- or 96-h average) < the lower 10th percentile of the distribution of the 48- or 96-h LC5 estimates from acute tests?
- Or, where chronic toxicity data are *required* or available, is the upper 10th percentile of EECs (21-, 60-, or 90-d average) < the lower 10th percentile distribution of the 21-, 60-, or 90-d NOECs from chronic tests?

These criteria would be equivalent to a 1% risk of any randomly selected species being affected at the LC5 or the chronic NOEC at the modeled site in one year of any given (future) ten-year period. The characterization may also allow some of the uses (*e.g.,* corn) modeled above to be excluded from the Tier 3 characterization. By excluding uses of little potential risk from further analysis, resources can be focused on those uses where the potential risks are greatest.

Tier 3: Toxicity. The same toxicity data as for Tier 2 would be used.

Tier 3: EEC. The EEC modeling should be focused on probability distributions from multiple sites within those uses that failed to pass Tier 2. Multiple scenarios should be modeled for each use to define geographical as well as climate-driven EEC distributions so that problem areas for each use can be identified. From 10 to 50 sites (representing each pesticide use identified in Tier 2 as having potential substantial risk) should be modeled over ≥30 years to give for each specified duration of exposure (*e.g.,* 4 h, 96 h, 21 d, etc.) a distribution of estimated maximum annual average dissolved EECs for each scenario modeled. When possible, all values of the $t^1/_2$ and K_{OC} should be used as model inputs via techniques like Monte Carlo. If not, use input values as in Tier 2. More realistic management practices should be used. The models should also be used at this point to identify the impact of mitigation such as rate reduction, formulation change, area restrictions, management practices, appropriate buffers, application frequency, etc. Weather data should be used to consider the 1-in-10-year maximum concentrations for the duration of interest. Each site/scenario should be modeled to provide a distribution (over years) of estimated annual maximum (specified duration) average dissolved EECs.

For each desired duration and each site/scenario modeled, the distribution of annual maximum average dissolved EECs should be plotted as a cumulative frequency exceedance curve. Each data point on any given EEC cumulative frequency exceedance curve for site j will have coordinates

y_{ij} = annual maximum (specified duration such as 48 h, 96 h, 21 d, 60 d, 90 d, etc.) average dissolved EEC for year i at site j and

x_{ij} = fraction or percentage of years at site j in which the annual maximum (specified duration) average dissolved EEC $\geq y_{ij}$.

These coordinates should be used to determine the upper 10% of years annual maximum average dissolved EEC for each site/scenario modeled.

Tier 3: Assessment. Tier 3 is conducted to identify those sites/scenarios of potential substantial risk in each use identified in Tier 2 as having substantial potential risk. This is done by comparison of the point estimates of the upper 10% annual EECs to the lower 10th percentile of the toxicity distribution. This can be represented in a series of questions for each site/scenario:

- Is the upper 10th percentile of the EECs (for a particular site/scenario and 48- or 96-h average) < the lower 10th percentile of the distribution of the 48- or 96-h LC5 estimates from acute tests?
- Or, where chronic toxicity data are available, is the upper 10th percentile of the EECs (21-, 60-, or 90-d average) < the lower 10th percentile distribution of the 21-, 60-, or 90-d NOECs from chronic tests?

These criteria would be equivalent to a 1% risk of any randomly selected species being affected at the LC5 or the chronic NOEC at the modeled site in one year of any given (future) ten-year period. The characterization may allow some of the modeled sites/scenarios for a given use (*e.g.*, cotton grown in heavy soil under irrigation) to be excluded from mitigation or further characterization. This would again focus resources on those sites/scenarios where problems would be most likely to occur.

Tier 4: Toxicity. It is recommended that more realistic toxicity tests be used at this stage of the assessment. The regulator may require or the registrant may furnish additional (field-derived) toxicity data that are more directed, or special tests for pulse exposures of pesticides (such as those with very short half-lives) may be used to obtain a better estimate of the toxicity distribution.

Tier 4: EEC. Although the modeling methods are not yet completely available, it is recommended that characterizations of the EEC be based on landscape modeling (which takes into account the location of treated areas with respect to surface waters and percentages of land treated with a particular chemical).

At this stage, it is applicable to consider the impact of landscape factors on the likelihood of experiencing the "edge-of-field" EECs estimated in Tiers 2 and 3. Landscape factors do not always reduce the maximum EEC value possible. However, they provide the risk assessor with better estimates of how often certain exposure levels might occur. Landscape factors usually reduce the mean "edge-of-field" modeled EEC values significantly.

There are currently no all-encompassing simulation models that allow the simultaneous consideration of all landscape factors. Instead, the impact of each factor on the distribution of EEC values may have to be examined separately or, for spatial factors, in combination using GIS. Many other combinations of the above variables might also prove valuable.

Tier 4: Assessment. The criteria would be very similar to those used for Tier 3. Tier 4 is conducted to help develop more realistic aquatic risk assessments for those pesticides, uses, and sites/scenarios identified in the lower tiers as having potential substantial risk. The use of landscape factors will generally reduce the apparent nationwide frequency of occurrence of higher EEC values and will also permit identification of sites/ scenarios most in need of mitigation and/or monitoring. For these reasons, landscape factors should when practical be considered before the need for mitigation or available mitigation alternatives are considered.

Certain landscape factors may be product specific; however, some generic factors are the following:

- percentage of use area treated with pesticide of interest (by state or county)
- spatial distribution of potentially erosive soil/slope combinations adjacent to water bodies
- spatial distribution of cropped and non-cropped land within watersheds or adjacent to ponds
- extent and nature of natural buffers between use sites and water bodies
- spatial distribution of rates or timings of chemical application
- spatial distribution of wind speeds and directions or other meteorological factors (for drift issues)
- spatial distribution of adopted mitigative measures (*e.g.*, soil conservation practices)
- spatial distribution of application methodologies
- distribution of water bodies of different types and ecological significance in the region of interest

While this approach has advantages over the quotient method of risk assessment, there are a number of limitations (see Sections 1.1.4 and Chapter 5). These areas represent opportunities for methods improvement in the future:

- In spite of widespread use and acceptance of the species sensitivity distributions for water quality criteria and for regulation by the government of the Netherlands, there has been some criticism of the statistical and biological assumptions associated with this approach (Smith and Cairns, 1993).
- The choice of a level protective of 90% of species may not be acceptable if the 10% of potentially affected species includes endangered species or organisms of high ecological, commercial, or recreational significance.
- Indirect effects such as effects on predator-prey relationships or food chain bioaccumulation are not addressed. Effects of pesticide-contaminated sediments on sediment-dwelling organisms are not presently determined. Ecosystem-level effects (*e.g.*, effects on nutrient cycling or organic matter processing) are not addressed.
- Lack of measurement of indirect and ecosystem-level effects could be problematic when combined with a lack of biological monitoring data to support mitigation effectiveness. Given that many new pesticides are much less toxic to fish than to aquatic invertebrates, reliance on "fish kills" as evidence of insufficient pesticide mitigation may be inadequate (Mayer and Ellersieck, 1986).

- Long-term effects associated with multiple-year exposure or cumulative effects resulting from multiple pesticides or other stressors are not addressed.
- Proposed assessment and measurement endpoints may not be appropriate for evaluating effects associated with larger spatial scales (*e.g.*, landscape or watershed levels).

2.4 TESTING THE MODELS

2.4.1 Preliminary Evaluation of LOCs with Incident and Experimental Data from the Field

As was suggested earlier, it may be possible to validate current LOCs through the use of incident data from field observations that are linked to measurements of actual environmental concentrations (AECs). The literature has a number of reports of fish and crustacean kills associated with pesticides entering waters. Seven studies and a portion of easily retrievable incident data in OPP were used in the following analysis. These are summarized in Table 2-3.

TABLE 2-3. Toxicity from Incident and Field Experimental Studies on Pesticides in Aquatic Species

Organism	Endpoint	No-Observed-Effect Concentration (NOEC) in µg/L	Actual Environmental Concentration (AEC) in µg/L	LC50 in µg/L	Ratio of AEC and LC50
FIELD STUDY DATA (Eaton et al., 1985; Brazner et al., 1988; Knuth et al., 1992; Springborn, 1989; Lozano et al., 1989; Webber et al., 1992; Fairchild et al., 1992)					
		Chlorpyrifos			
Fathead minnow	Spinal deformity (19-24%)	NA	5.7	203	0.03
Bluegill	38% mortality	NA	6.3	1.3	4.85
Bluegill	1.7% mortality	NA	0.51	1.3	0.39
Fathead minnow	0.8% mortality	6.3	32	203	0.16
Cladocera	Significant reduction in numbers	NA	0.51	1.7	0.30
		Azinphos-methyl			
Cladocera	Significant reduction in numbers	4.1	20.4	1.1	18
Bluegill	Lack of reproductive effects	4.0		4.1	
Zooplankton	LOEL	NA	5.2	1.1	4.73
Bluegill adults	Significant reduction in survival	0.2	5.2	4.1	1.27
		Fenvalerate			
Cladocera	Significant population reduction	0.01[a]	0.047	0.05	0.94
Bluegill	Reduced growth	0.01[a]	0.08	0.26	0.31
Bluegill	Significant mortality	0.2	0.92	0.26	3.54
Bluegill	Decline in biomass	0.01	0.18[b]	0.50	0.36
Bluegill	Reduced survival and biomass production	0.27	0.77	0.26	2.96
Bluegill	Reduced reproductive success	NA	0.27	0.26	1.04
Micro-invertebrates	Significant reduction in numbers	0.01	0.18[b]	0.05	3.60

a = Nominal concentration

b = Water collected one day after application

TABLE 2-3. (continued)

Organism	Endpoint	No-Observed-Effect Concentration (NOEC) in μg/L	Actual Environmental Concentration (AEC) in μg/L	LC50 in μg/L	Ratio of AEC and LC50
INCIDENT DATA (Urban, 1993; Scott et al., 1989)					
Chlorpyrifos					
Bluegill	Kill	NA	2800	1.3	2153
Azinphos-methyl					
Striped mullet	Significant mortality	NA	3.72	0.38	9.79
Grass shrimp	30% mortality	NA	0.104	1.3	0.08
Endosulfan					
Spot	Kill	NA	0.14	0.09	1.56
Spot	Kill	NA	0.93	0.09	10.33
Spot	Kill	NA	0.73	0.09	8.11
Striped mullet	Kill	NA	0.14	0.38	0.37
Striped mullet	Kill	NA	1.2	0.38	3.16
Blue crab	Kill	NA	0.14	19	0.01
Blue crab	Kill	NA	1.2	19	0.06
Blue crab	Kill	NA	0.73	19	0.04
Mullet	Kill	NA	0.73	0.38	1.92
Bluegill	Kill	NA	0.1	1.2	0.08
Fenvalerate					
Striped mullet	Significant mortality	NA	0.107	0.58	0.18

Unfortunately, the paucity of published data from field studies and the lack of consistency in the sampling and analysis of the matrix for pesticide residues during field incidents do not allow meaningful conclusions to be drawn from the data. The ratios of the AEC/LC50 in the field studies (where one would expect more realistic characterization of effects and concentrations at which these were observed) were not consistent, and it was not possible to confirm a lack of response at concentrations similar to low LC values. Obviously, a larger data set with better information on actual exposure concentrations would have to be used to better evaluate such a relationship. This data set also points out some of the potential problems that may be associated with the use of field monitoring to assess ecosystem effects of pesticides, or for that matter, the reduction in effects that may result from mitigation measures. Careful study design can alleviate some of the problems.

2.4.2 Distributions of Susceptibility

The susceptibility data in the USEPA data base (Urban, 1993) is limited in its scope and range of organisms, but it has been developed from recent studies (conducted under GLP) and has been peer reviewed. A subset of the available data was used to assess the validity of some of the assumptions of distribution of toxicity data and the EEC. Although only small numbers of species were tested, it is possible to plot these data in rank order to assess the slope of the susceptibility distribution. Results are shown in Figure 2-11. The major difficulty in this analysis is a lack of knowledge of the appropriateness of the species used in testing. Atrazine (Figure 2-11) was the only pesticide for which susceptibility data were available for plants (mainly algae). The data show a reasonably good relationship between rank and LC50 or EC50 value. However, this relationship is not as good for those pesticides for which few data are available (for example, cyfluthrin).

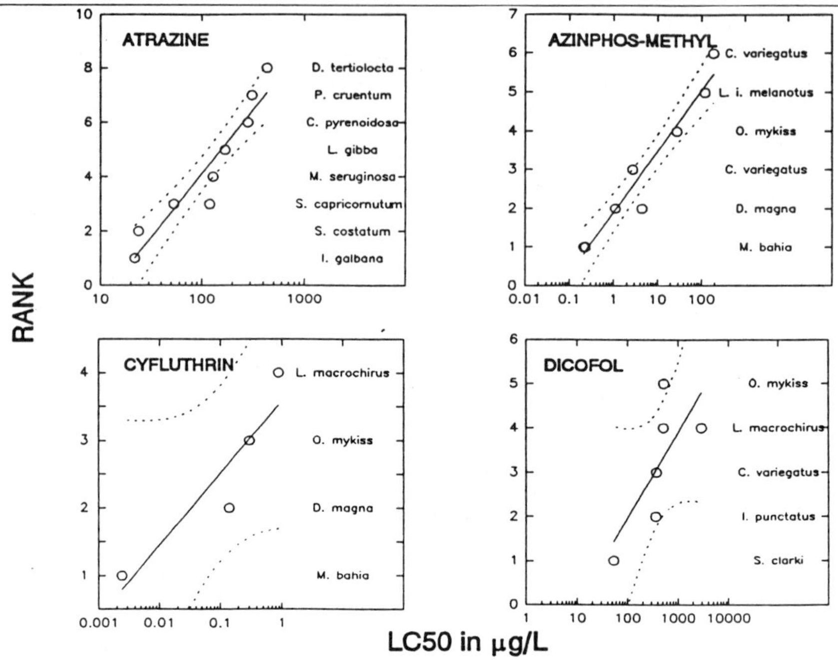

Figure 2-11. The relationship between rank order of toxicity and LC50 concentration in aquatic species for several pesticides (dotted lines show the 95% confidence intervals of the regression lines)

Other published sets of susceptibility data were available, and these were also tested. The toxicity of azinphos-methyl and chlorpyrifos has been reviewed (Odenkirchen and Eisler, 1988; Mayer and Ellersieck, 1986). In this case, the rank order of the susceptibility data was transformed to percent of the total rank order. On the assumption that these ranks more closely represented the "universe" of species (which are assumed to be normally distributed with respect to rank), a probability transformation was used on the rank orders (Figure 2-12). The data for chlorpyrifos show a reasonable fit to the model relationship; however, the data for azinphos-methyl show a poorer fit to the model. It is interesting to note that the range over which the toxicities were observed is very similar for both pesticides. The fact that chlorpyrifos and azinphos-methyl are both organophosphorous pesticides with similar mechanisms of action and similar toxicity may explain this observation. The lowest LC50 value for azinphos-methyl in the USEPA data base (0.2 µg/L) was close to that observed in the larger data set (0.1 µg/L). The USEPA data base only had two data points for chlorpyrifos (in crustacea), the lowest of which was 10.2 µg/L. If this were the only data set available, it would have seriously underestimated the lowest toxicity observed in the larger data set (Figure 2-12).

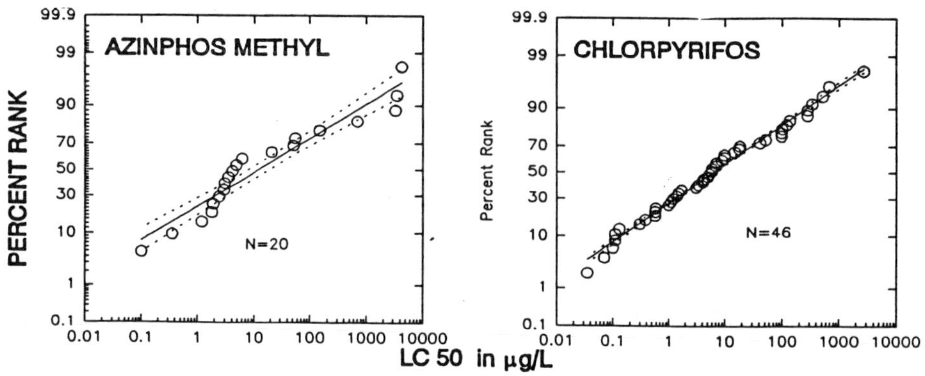

Figure 2-12. Plots of LC50 in aquatic species versus percent rank order for azinphos-methyl and chlorpyrifos (dotted lines show the 95% confidence intervals of the regression line)

2.4.3 Comparing Toxicity Data to AECs

The probabilistic risk assessment procedure is based on a comparison of EEC and toxicity data generated in laboratory studies. Validation of this could be carried out by comparison of AECs to a distribution of toxicities. Although an extensive search of the literature was not conducted, little data were found for AECs of pesticides. However, atrazine levels in surface waters in Ontario and Quebec have been surveyed over a period of several years (Duval and Gauthier, 1981; Frank *et al.*, 1979; Muir *et al.*, 1978; Roberts *et al.*, 1979), and these data were available in electronic form. This data set consisted of about 900 measurements, some of which were below the level of detection (LOD = 0.02 µg/L). The data were transformed via an uncensoring procedure (Newman *et al.*, 1992) to generate estimates of the data points below the LOD. The frequency distribution of these data was plotted against the AEC and compared to the toxicity data for atrazine in aquatic plants. Although there was a small overlap between the distribution of the AEC and toxicity (Figure 2-13), a considerable margin of safety existed between these lines if the upper 95 percentile of the concentration distribution was taken into consideration. The toxicity tests were short term (5 d); however, the life cycle of these test organisms (except for duckweed) is such that this length of exposure is equivalent to a multi-life-cycle test with chronic exposures. In addition, the data were obtained over short sampling periods. In actual fact, concentrations averaged over the period of the toxicity tests (5 d) would probably be lower. On this basis, the application of this model to real-world data would suggest that mitigation for the effects of atrazine in surface waters under the conditions of use in Ontario and Quebec would not be necessary.

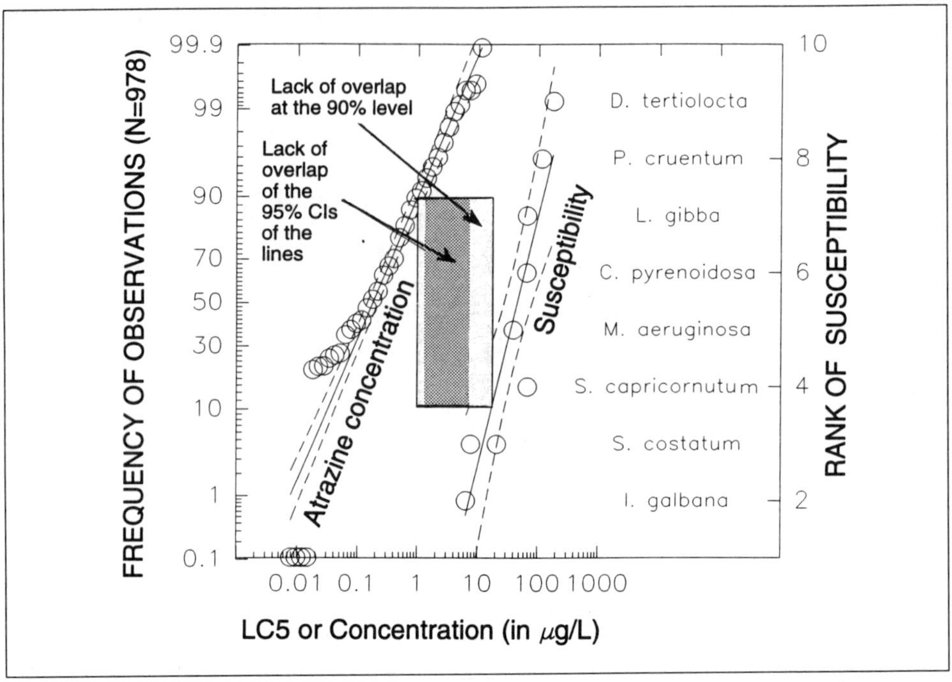

Figure 2-13. The relationship between environmental concentrations of atrazine measured in Ontario and Quebec (1977-1989) and toxicity to aquatic plants (algae)

2.4.4 Testing the Assumptions in the Tier 1 LOCs

The criteria for testing the assumptions of Tier 1 were

- a single-event EEC based either on a high-exposure scenario resulting from direct overspray of a pond 2 m deep or 10% runoff from a 10 ha field into a 1 ha pond and
- acute toxicity for the most sensitive species expressed as the LC5, LC1, LC0.1, and LC0.01.

This EEC is very conservative. Using the USEPA data base (Urban, 1993), it was possible to test acute LC values calculated from the slopes of the concentration-response lines and to compare these to EECs calculated from the highest recommended label rate for the pesticide (Thompson, 1985a; 1985b; 1985c). These comparisons were restricted to those data base entries for which at least three entries were available (those entries are also equivalent to the current USEPA acute toxicity information [two fish, one crustacean]).

These results (Figure 2-14) suggest that, where the very conservative LC0.01 is taken as the toxicity criterion, about 89% of the pesticides would pass on to Tier 2 of the assessment scheme. If the criterion of the LC5 were used, 75% of the pesticides would pass to Tier 2. In this case, those pesticides that would be passed on to Tier 2 include those pesticides for which field-incident data have been reported (such as azinphos-methyl). Pesticides that would not pass to Tier 2 do not include pesticides that have been associated with field incidents. These results are based on a small sample of data and should be tested with a larger data set that includes more pesticides known to be associated with incidents in the field (endosulfan, etc.).

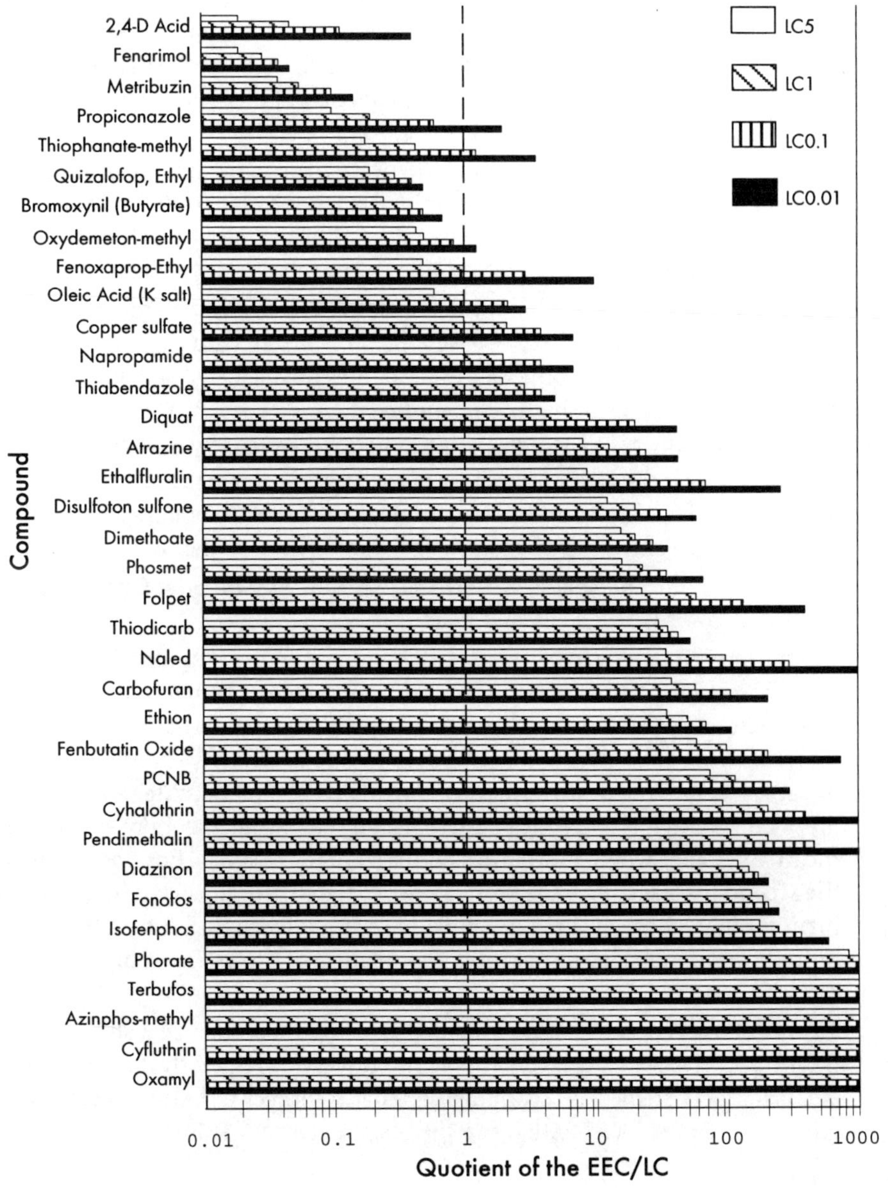

Figure 2-14. The quotient of an EEC calculated for Tier 1 and various LC values (based on data from Urban, 1993, and Thompson, 1985a, 1985b, 1985c)

2.5 RECOMMENDATIONS

2.5.1 Risk Assessment Recommendations

Short Term

- Develop a detailed tiered approach to aquatic risk assessment that shows how the various possible tiers of EECs, runoff studies, and surface water monitoring studies are used in the risk assessment. These would begin with a simple assessment using a conservative criterion for judgment. As the assessment is refined with more likely estimates of exposure and EECs, less conservative and more realistic criteria can be used.
- For Tier 1 of the assessment, use the following toxicity data: acute dose responses in three freshwater aquatic animals and one plant (and in some cases, saltwater organisms), half-life in water, K_{ow}, and in some cases, chronic toxicity dose responses in four species of aquatic organisms.
- In selecting a toxicity criterion for determining LOCs, it is recommended that the LC5 be used. A retrospective assessment of current data suggests that this criterion would capture some 75% of pesticides for Tier 2. A validation study on the selection of the LC5 should be conducted.
- Conduct a detailed analysis of existing field incident data and results of field studies to assess the appropriateness of the trigger values (LOCs) currently used and proposed for use. In addition, conduct a retrospective analysis of existing field study data and apply to assessment of mitigation requirements.
- Include temporal and spatial patterns of exposure and broader representation of resources at risk in risk assessments.
- Use the temporal pattern of mortalities in laboratory toxicity studies in refined risk assessments. Consider two scenarios from a flow-through constant dosing study: one where all mortalities occurred within the first day and one where the mortalities occurred over the course of the study and were weighted more towards day 4. If the pesticide is very short-lived in aquatic systems (*e.g.*, half-life = 1 day) and it is used once per year, then there would be more concern in the first instance where mortalities occurred in the first day, but not as much in the second instance since the pesticide would not naturally occur in the environment for that length of time. A follow-up static study could confirm the lack of mortality and, thus, lower risk after degradation of the pesticide in a pulse-dose scenario.
- Timing of use can also have an impact on the type of studies that should be performed. If a pesticide is used only once or very infrequently during the year and is not persistent, it is recommended that resources be better focused on additional acute studies with more species than chronic studies.

Long Term

- Employ GIS to more fully understand the actual use of a pesticide and its relation to valued ecological resources.

- Temporal and spatial patterns of exposure risk assessments should include information on the regional distribution, timing of use, and extent of use of the pesticide.

2.5.2 Hazard Assessment Recommendations

Short Term

- Shift resources expended on fish life-cycle study to chronic studies on fish early life stages and chronic invertebrate studies.
- Use the following triggers for chronic studies:
    - The pesticide is intended for use such that its presence in water is likely to be continuous or recurrent,
    - the EEC $\geq$ 0.01 of the LC50 or EC50, and
    - the half-life in water is greater than 4 days or the physicochemical properties indicate cumulative effects.

Long Term

- Trigger tests of specific life stages on effects observed in early life-stage studies or from results of mammalian or avian reproductive tests.
- Develop alternatives to the fish life-cycle study that target specific life stages or functions.
- Express the results of a refined hazard analysis as a distribution of toxicity values rather than a single point estimate. A major advantage of this approach is that it uses all relevant single-species toxicity data, and when combined with exposure distributions, allows quantitative estimates of risks to aquatic organisms.
- Use customized tests to simulate actual use patterns. For example, if the pesticide in question is used once per year and has a half-life in water of 1 hour, then it would be beneficial and more realistic to consider a static test system instead of a continuous-flow constant concentration test. Similarly, if the pesticide is used once a week for a number of weeks, then it would be more appropriate to utilize a pulse-dosed system to mimic the incremental inputs of the pesticide over time.
- Consider the impact on sediment-dwelling species and develop acceptable protocols for performing a core set of studies.
- Assess bioavailability of pesticides in sediments in combination with toxicological effects data to provide information about the risk associated with chemical/ sediment interactions.

REFERENCES

Birge, W.J. and J.A. Black. 1982. Statement on surrogate species cluster concept. Report of the Surrogate Species Workshop, Life Systems, Cleveland, OH, pp. A6-5 - A6-7.

Blanck, H., G. Wallin and S.-A. Wangberg. 1984. Species dependent variation in algal sensitivity to chemical compounds. *Ecotoxicol. Environ. Safety* 8:339-351.

Brazner, J.C., S.J. Lozano, M.L. Knuth, S.L. Bertelsen, L.J. Heinis, D.A. Jensen, E.R. Kline, S.J. O'Halloran, K.W. Sargent, D.K. Tanner and R.E. Siefert. 1988. The effects of chlorpyrifos on a natural aquatic ecosystem: A research design for littoral studies and final research report. USEPA ERL-Duluth, Duluth, MN.

Cardwell, R.D., B.R. Parkhurst, W. Warren-Hicks and J.S. Volosin. 1993. Aquatic ecological risk. *Water Environment and Technology* 5:47-51.

CEPA. 1992. Canadian Environmental Assessment Act (CEAA). Third Session, Thirty-fourth Parliament of Canada, 40-41 Elizabeth II, 1991-1992, pp. 1-54.

Crossland, N.O., F. Heimback, I.R. Hill, A. Boudou, P. Leeuwangh, P. Matthiessen and G. Persoone. 1992. Summary and recommendations of the European Workshop on Freshwater Field Tests (EWOFFT). Potsdam, Germany, June 25-26, 1992.

Delon, C., W. Banks, K. Barylski, R. April, D. Moon and J. Romney. 1991. Water quality criteria and standards. In Technical support document for water quality-based toxics control. EPA/505/2-90-001 PB91-127415. USEPA Office of Water, Washington, DC, pp. 29-46.

Doherty, F.G. 1983. Interspecies correlations of acute aquatic median lethal concentration for four standard testing species. *Environ. Sci. Technol.* 17:661-665.

Duval, D. and R. Gauthier. 1981. Présence des herbicides dans le fleuve saint-laurent et ses tributaires. Environment Canada, Inland Waters Directorate, Water Quality Program, pp. 1-33.

Eaton, J., J. Arthur, R. Hermanutz, R. Keifer, L. Mueller, R. Anderson, R. Erickson, B. Nordling, J. Rogers and H. Prichard. 1985. Biological effects of continuous and intermittent dosing of outdoor experimental streams with chlorpyrifos. In R.C. Bahner and D.J. Hansen, eds., *Aquatic Toxicology and Hazard Assessment: Eights Symposium*, ASTM STP 891. American Society for Testing and Materials, Philadelphia, PA, pp. 85-118.

Fairchild, J.F., T.W. LaPoint, J.L. Zajicek, M.K. Nelson, F.J. Dwyer and P.A. Lovely. 1992. Population-, community-, and ecosystem-level responses of aquatic mesocosms to pulsed doses of a pyrethroid insecticide. *Environmental Toxicology and Chemistry* 11:115-129.

Frank, R., G.J. Sirons, R.L. Thomas and K. McMillan. 1979. Triazine residues in suspended solids (1974-1976) and water (1977) from the mouths of Canadian streams flowing into the Great Lakes. *Journal Great Lakes Research* 5:131-138.

Giesy, J.P. and R.L. Graney. 1989. Recent developments in and intercomparisons of acute and chronic bioassays and bioindicators. *Hydrobiologia* 188/189:21-60.

GLWQG. 1993. Great Lakes Water Quality Guidelines, unofficial prepublication copy, March 31, 1993.

Health Council of the Netherlands. 1993. Ecotoxicological risk assessment and policy-making in the Netherlands - dealing with uncertainties. *Network* 6(3)/7(1):8-11.

Kenaga, E.E. 1978. Test organisms and methods useful for early assessment of acute toxicity of chemicals. *Environ. Sci. Technol.* 12:1322-1329.

Kimerle, R.A., A.F. Werner and W.J. Adams. 1985. Aquatic hazard evaluation principles applied to the development of water quality criteria. In R.D. Cardwell, R. Purdy and R.D.

Bahner, eds., *Aquatic Toxicology and Hazard Assessment.* Am. Soc. Testing and Materials STP 854. ASTM, Philadelphia, PA, pp. 538-547.

Knuth, M.L., L.E. Anderson, L.J. Heinis, R.R. Hicks, A.W. Jarvinen, S.J. Lozano, S.L. O'Halloran, J.A. Patzer, R.W. Seelke, R.E. Siefert, M.E. Sierszen, F.S. Stay and D.K. Tanner. 1992. Effects, persistence and distribution of azinphos-methyl in littoral enclosures. USEPA, ERL-Duluth, MN.

Lozano, S.J., J.C. Brazner, M.L. Knuth, L.J. Heinis, K.W. Sargent, D.K. Tanner, L.E. Anderson, S.L. O'Halloran, S.L. Bertelsen, D.A. Jensen, E.R. Kline, M.D. Balcer, F.S. Stay and R.E. Siefert. 1989. Effects, persistence and distribution of esfenvalerate in littoral enclosures. USEPA, ERL-Duluth, MN.

Mayer, F.L., Jr. and M.R. Ellersieck. 1986. *Manual of Acute Toxicity: Interpretation and Data Base for 410 Chemicals and 66 Species of Freshwater Animals.* U.S. Fish and Wildlife Service, Pln. # 160, Washington, DC.

Mount, D. 1982. Rationale for surrogate cluster concept. Report of the Surrogate Species Workshop, Life Systems, Cleveland, OH, pp. A6-1 - A67-4.

Muir, D.C.G., J.Y. Yoo and B.E. Baker. 1978. Residues of atrazine and N-deethylated atrazine in water from five agricultural watersheds in Quebec. *Archives of Environmental Contamination and Toxicology* 7:221-235.

Newman, M.C., D. Greene, P.M. Dixon, B.B. Looney and C. Segal. 1992. UNCENSOR v 3.0. SREL-44 UC-66e. Savannah River Ecology Laboratory, Aiken, SC.

National Research Council [NRC]. 1993. *Issues in Risk Assessment.* National Academy Press, Washington, DC.

Odenkirchen, E.W. and R. Eisler. 1988. Chlorpyrifos hazards to fish, wildlife and invertebrates: A synoptic review. U.S. Fish and Wildlife, Biological Report 85(1.13).

Okkerman, P.C., E.J. v. d. Plassche, H.J.B. Emans and J.H. Canton. 1993. Validation of some extrapolation methods with toxicity data derived from multiple species experiments. *Ecotoxicology and Environmental Safety* 25:341-359.

Roberts, G.C., G.J. Sirons, R. Frank and H.E. Collins. 1979. Triazine residues in a watershed in Southwestern Ontario (1973-75). *Journal of Great Lakes Research* 5:246-255.

Scott, G.I., D.W. Moore, M.H. Fulton, T.W. Hampton, J.M. Marcus, G.T. Chandler, D.S. Jackson, A.H. Baughman, L. Williams, C.J. Louden and E.R. Patterson. 1989. Agricultural insecticide runoff effects on estuarine organisms: Correlating laboratory and field toxicity testing with ecotoxicological biomonitoring. USEPA, ERL-Gulf Breeze, FL.

SETAC. 1994. *Life-Cycle Assessment Data Quality: A Conceptual Framework.* Report of the workshop held October 4–9, 1992 at Wintergreen, VA. Society of Environmental Toxicology and Chemistry, Pensacola, FL.

SETAC-Europe. 1991. Guidance Document on Testing Procedures for Pesticides on Freshwater Mesocosms. From the workshop "A Meeting of Experts on Guidelines for Static Field Mesocosm Tests" held at Monks Wood Experimental Station, Abbotts Ripton, Huntingdon, UK, July 3–4, 1991.

Sloof, W., J.H. Canton and J.L.M. Hermans. 1983. Comparison of the susceptibility of 22 freshwater species to 15 chemical compounds. I. (Sub) acute toxicity tests. *Aquatic Toxicol.* 4:113-128.

Sloof, W., J.A.M. van Oers and D. de Zwart. 1986. Margins of uncertainty in ecological hazard assessment. *Environmental Toxicology and Chemistry* 5:841-852.

Smith, E.P. and J. Cairns, Jr. 1993. Extrapolation methods for setting ecological standards for water quality: statistical and ecological concerns. *Ecotoxicology* 2:203-219.

Springborn Laboratories. 1989. Report # 89-06-2935. Wareham, MA.

Stephan, C.E., D.I. Mount, D.J. Hansen, J.H. Gentile, G.A. Chapman and W.A. Brungs. 1985. Guidelines for deriving numerical national water quality criteria for the protection of aquatic organisms and their uses. NITS PB85-227049. USEPA Research Laboratory, Duluth, MN, Narragansett, RI, and Corvallis, OR.

Thompson, W.T. 1985a. *Agricultural Chemicals Book I Insecticides.* Thompson Publications, Fresno, CA.

Thompson, W.T. 1985b. *Agricultural Chemicals Book II Herbicides.* Thompson Publications, Fresno, CA.

Thompson, W.T. 1985c. *Agricultural Chemicals Book IV Fungicides.* Thompson Publications, Fresno, CA.

Urban, D. 1993. D. Urban, USEPA, personal communication.

USEPA. 1992. *Framework for Ecological Risk Assessment.* EPA/630/R-92/001. Washington, DC.

USEPA. 1984. *Federal Register* 49:(207), October 24, 1984, pp. 42895. Washington, DC.

Webber, E.F., W.G. Deutsch, D.R. Bayne and W.E. Seesock. 1992. Ecosystem-level testing of a synthetic pyrethroid insecticide in aquatic mesocosms. *Environmental Toxicology and Chemistry* 11:87-105.

Chapter 3: EXPOSURE CHARACTERIZATION

The estimation of exposure levels of aquatic organisms to pesticides is an important part of aquatic risk characterization. In this chapter, emphasis is placed on computer modeling rather than on monitoring estimates of expected environmental concentrations (EECs) of pesticides to be consistent with the **new paradigm** (see Introduction) of the Office of Pesticides Programs' (OPP) aquatic risk characterization. However, as will be discussed here and in the chapter on mitigation verification, field and monitoring studies will continue to play an important role in the characterization of aquatic exposures. Such studies help to validate computer modeling estimates, characterize actual pesticide concentrations in natural waters, and determine the effectiveness of mitigation methods in reducing the exposure of aquatic organisms to pesticides.

Prior to the new paradigm, EECs were developed in a variety of ways. Some were developed through actual measurement of water concentrations in field or monitoring studies, while some were derived from "back of the envelope" calculations with conservative assumptions concerning the proportions of applied pesticide directly entering water bodies. Other EECs were developed from "high to very high runoff" computer modeling sites/scenarios. However, despite differences in their method of development, all represented single EECs whose quantitative likelihood of occurrence was unknown, because they had not been placed in the probabilistic framework that is necessary to capture the high spatial and temporal variation of actual pesticide concentrations in the environment. At best, EECs could be only qualitatively described as "high to very high exposure." Single EECs are now used only in the preliminary tiers of the U.S. Environmental Protection Agency's (USEPA) new paradigm of aquatic risk characterization, where emphasis is placed on a continual refinement and mitigation of EEC levels.

The higher tiers of the new aquatic risk characterization paradigm call for the replacement of single EECs with temporal and spatial distributions of EECs that relate directly to the quantitative likelihood of their occurrence, so that levels of risk to the aquatic community can be quantitatively assessed and related to mitigation (if necessary). This chapter defines a series of approaches for estimating EEC distributions that always place individual EECs in a probabilistic framework. Single-valued estimates of EECs, like those from "high to very high runoff" modeling sites/scenarios, will be defined quantitatively on the basis of the probability of their occurrence. Such probabilistic EEC distributions for one to multiple sites over multiple years can be used to estimate where and how frequently the maximum or average concentrations of pesticides will exceed any given threshold for acute or chronic toxicity to aquatic organisms.

A Federal Insecticide, Fungicide and Rodenticide Act (FIFRA) Exposure Modeling Work Group (EMWG) comprised of academic, OPP, U.S. Department of Agriculture (USDA), and industry scientists is currently attempting to develop uniform methods for generating, documenting, presenting, and validating accurate and scientifically defensible pesticide EEC distributions for surface water in a reasonable time frame. As part of this effort, the workgroup has identified suitable models and supporting data bases for simulating pesticide runoff, leaching, and fate in surface and ground water. This document does not address the technical aspects involved in implementing adequate modeling and directs its attention to general approaches and recommendations to aid the modeling development efforts.

Specific topics in this chapter include a discussion of the inputs and outputs of primary and secondary models that simulate spray drift, pesticide runoff, and fate in surface water. There are discussions of random errors due to uncertainties in the model input, systematic errors due to deficiencies in the models, stochastic modeling, and model verification. There is also a discussion of natural EEC variability, its sources, and how it is handled through a probabilistic modeling approach. In addition, there is a section on the potential uses and limitations of modeling. Finally, there is a list of recommendations to risk assessors and developers of EEC data.

3.1 MODEL INPUTS, SUPPORTING DATA BASES, TYPES, AND OUTPUTS

Computer models can be used to simulate the environmental fate and transport of pesticides and in some cases their degradation products. Direct and/or indirect inputs to various types of such environmental fate models (*e.g.*, runoff, spray drift, and surface water as defined below) include the following:

- Meteorological Data—precipitation, temperature, solar irradiation, relative humidity, wind speed, wind direction
- Crop Characteristics and Agricultural Practices—dates of planting, emergence, maturity, and harvest; crop cover, foliar interception, post-harvest residue cover, root depth, soil incorporation depth, tillage practices, rotation practices, irrigation practices, spray elevation, spray nozzle sizes, droplet size spectrum; and number, dates, rates, and methods of pesticide applications
- Soil and/or Sediment Properties—organic content, pH, soil texture, initial soil moisture, field capacity, wilting point, saturated hydraulic conductivity, slope, temperature, bulk density and porosity as a function of vertical and/or horizontal segmentation
- Pesticide Properties—aqueous solubility, disassociation constant acid/base (pK_as), Henry's constant, octanol/water partition coefficient, aqueous diffusivity, organic normalized soil/water and/or sediment/water partition coefficients, foliar washoff and degradation, overall and process-specific half-lives, and degradate formation rate constants
- Surface Water Characteristics—pesticide loadings due to runoff, soil erosion, and spray drift; pH, temperature, suspended sediment concentrations, dissolved natural organic concentrations, redox potentials, base flows, depth, width, cross sectional areas, and dispersion coefficients

To ensure some uniformity in input by different users of computer models, the FIFRA EMWG has recommended to the scientific staff of OPP a set of primary data bases for satisfying the input data requirements for most FIFRA-related pesticide fate modeling.

Environmental fate and transport computer models can be classified according to the portion of the environment and/or the processes simulated by the model. The most common types of models for characterizing aquatic exposures, as well as some typical outputs from each type, are as follows:

- Dissolved Runoff, Soil Erosion, and Leaching Models—These simulate water and soil movement along with pesticide fate and transport to the edge of the field, and from the soil surface through the vadose zone to at least the bottom of the root zone. Outputs include event, monthly, and annual runoff volumes and sediment yields along with their associated dissolved and adsorbed pesticide loadings to surface waters.
- Spray Drift Models—These simulate the atmospheric transport and deposition of pesticide containing droplets that are generated during spraying of liquid pesticide formulations. Outputs include the pesticide deposition flux as a function of downwind distance from the application swaths from which spray drift pesticide loadings to surface waters can be computed.
- Surface Water Models—These simulate water flow, sediment deposition, and pesticide fate and transport in surface waters that receive dissolved runoff, soil erosion, and spray drift pesticide loadings. Outputs include dissolved pesticide concentrations in the water columns, pesticide concentrations adsorbed to suspended sediment, and pesticide concentrations adsorbed to the bottom sediment as a function of time and location (segment).

Dissolved concentration outputs can include initial, daily, maximum 48-hour (h) average, maximum 96-h average, maximum 21-day (d) average, maximum 60-d average, and maximum 90-d average concentrations for comparison to various water column acute and chronic aquatic toxicity thresholds. With additional coding, maximum average concentrations of any user-specified duration can be obtained. Daily and average monthly dissolved concentrations can be plotted as a function of time for a given pond or flowing water segment, along with one or more horizontal lines representing various toxicity thresholds (*e.g.*, a 48- and 96-h LC50 or a 21-, 30-, and 60-d MATC). Such plots can show how long and how frequently dissolved concentrations exceed the various horizontal aquatic toxicity thresholds (Figure 3-1). Daily or monthly concentrations can also be plotted as a function of location (segment) at a given time (*e.g.*, at peak runoff).

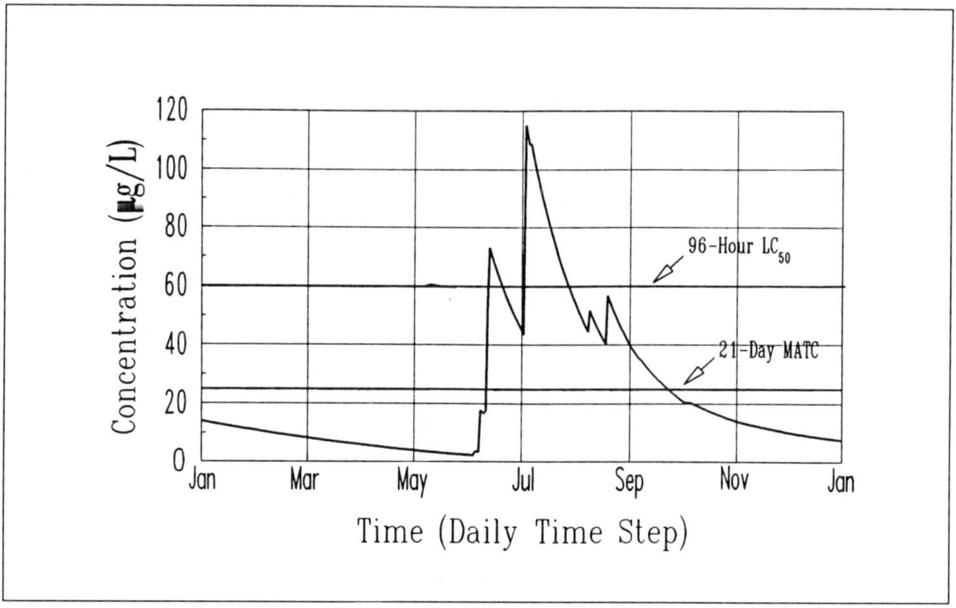

Figure 3-1. Daily EECs versus time compared to a toxicity threshold such as a 96-hour LC50 or 21-day MATC

Pesticide loading outputs from the runoff and spray drift models serve as inputs to the receiving water models (Figure 3-2).

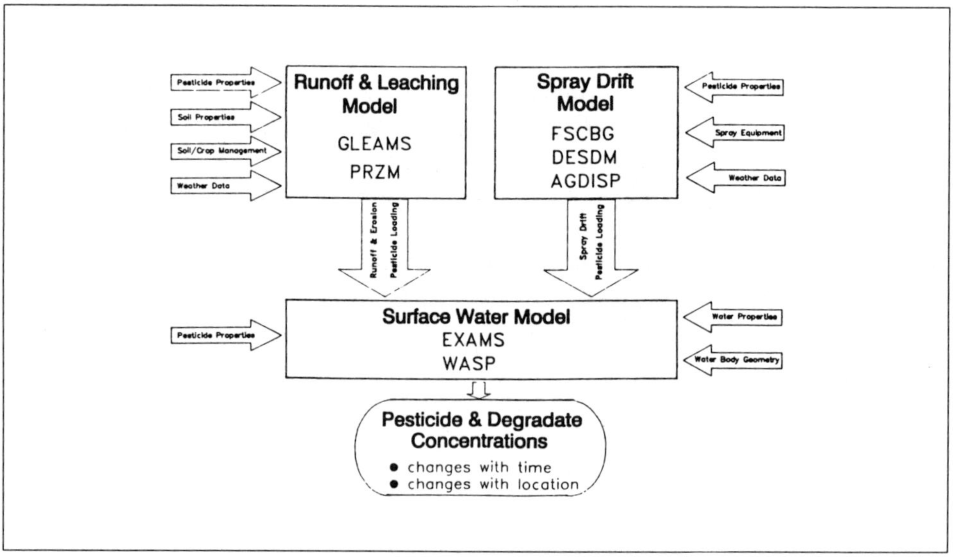

Figure 3-2. Inputs and outputs from pesticide fate and transport models

3.2 MODEL COMPARISONS

To support modeling uniformity, the FIFRA EMWG has recommended to the scientific staff of OPP Groundwater Loading Effects of Agricultural Management Systems (GLEAMS) and Pesticide Root Zone Model (PRZM2) as primary runoff/soil erosion models, and Exposure Analysis Modeling System (EXAMS) and Water Quality Analysis Simulation Program (WASP5) as primary surface water models. The FIFRA EMWG also recommends such models as Simulator for Water Resources in Rural Basins (SWRRB), Erosion Productivity Impact Calculator (EPIC), and Hydrological Simulation Program-Fortran (HSPF) as secondary models that might be more appropriate than the primary models in some cases. Generally, these models have been developed and applied to agricultural settings.

The two recommended primary runoff/soil erosion models (GLEAMS and PRZM2) share the following characteristics:

- Both are field, daily time scale models that can be used to estimate runoff, leaching, and associated pesticide loadings.
- Both use a similar modified SCS curve number method for estimating runoff and a "tipping bucket" storage routing method for predicting infiltration within the root zone.
- Both allow for different inputs of soil variable and lumped soil decay constant values for different soil layers and different inputs of the soil erosion cropping (C) and management (P) factors over time.
- Both simulate to some degree foliar as well as soil degradation, foliar interception, foliar washoff, plant uptake, crop rotations, daughter products, and some types of pesticide application, tillage, and irrigation methods.
- Both use empirical erosion/sedimentation technology; neither simulates subsurface lateral flow, macropore or preferential flow, tile drainage, nonlinear binding isotherms, adsorption/desorption kinetics, banded applications, formulations, turf, or orchards.

The two primary surface water models (EXAMS and WASP5) share the following characteristics:

- Both can accept multiple pesticide loadings into multiple stream segments.
- Both give pesticide concentrations in the water column and benthic sediment as a function of time for a given stream segment or as a function of stream segment at a given time.
- Both have identical chemistry/environmental fate algorithms which were originally developed for EXAMS. However, WASP5 has more sophisticated hydrology and sediment algorithms and can simulate non-steady state conditions.
- Both can simulate some specific dissipation processes including hydrolysis, direct photolysis, oxidation/reduction, biodegradation, and volatilization.

- Both originated in and are supported by the USEPA Laboratory in Athens, Georgia.

SWRRB, EPIC, HSPF, and Spray Drift Models. SWRRB is a basin-scale model that can address multiple climates, soils, and management practices within a watershed. EPIC was originally developed to simulate the effects of soil erosion on productivity and, as such, has some relatively sophisticated soil erosion and plant growth algorithms that neither PRZM2 nor GLEAMS has. The fairly recent incorporation of pesticide fate algorithms from GLEAMS also gives EPIC the capability of simulating pesticide losses due to runoff and soil erosion.

HSPF is a basin-scale model that can be used to simulate time varying runoff and soil erosion processes within a single storm event, which requires less than the minimum daily time input data. Since HSPF also includes a surface water model, it has the added advantage of not requiring an external interface with a surface water model. However, the surface water portion of HSPF is less detailed than WASP5.

Part of the above discussion on model comparisons was based upon the final report of the Aquatic Effects Dialogue Group (Wauchope, 1992). Products of the current FIFRA modeling workgroup include matrix comparisons of the recommended primary and secondary models as well as more detailed descriptions of the strengths and limitations of each model.

Based in part upon the work of the Spray Drift Task Force (SDTF), OPP will make recommendations with respect to spray drift models. The SDTF is evaluating two aerial spray drift models: AGDISP (a derivative for crop uses of the Forest Service Spray Drift Model) and Dow-Elanco Spray Drift Model (DESDM) developed by DowElanco. The SDTF is also evaluating an air blast orchard spray model currently being developed in Britain. The possibility of incorporating a stochastic component in the spray drift models is also being considered.

3.3 RANDOM AND SYSTEMATIC ERRORS IN MODELING RESULTS

Errors or uncertainties in the computer estimates of pesticide concentrations at a specific location and time frame within the environment can be classified as either random or systematic. Random errors are just as likely to overestimate the concentration as underestimate it. Systematic errors lead to routine overestimates or routine underestimates.

Random errors in the modeling estimates of pesticide concentrations are caused by the propagation of uncertainties in the inputs to the computer model. Such uncertainties are often due to

- inadequate data bases,
- uncertainties in the experimental values used for modeling input,
- the use of snapshot input values for values that randomly vary over time and/or space, or

- the use of fate input values derived under some set of experimental conditions that significantly differ from the set of environmental conditions being modeled.

Systematic errors are caused by deficiencies in the computer models. Such deficiencies include the assignment of inadequate spatial and/or time dimensions for spatially and/or time-variant input variables (*e.g.*, inability to assign different values of an input variable for different segments/soil layers, at the same time or for the same segment/soil layer, at different times within the same computer run). However, it should be noted that disaggregated models are not always more accurate than more aggregated models. Other deficiencies include failure of the various models to accurately simulate or simulate at all one or more important elements affecting the output, including complex hydrology, various dissipation pathways, formulations, management practices, proposed mitigation methods, or crop growth, uptake, interception, or washoff.

It is often easier to develop appropriate models than to collect the parameters necessary to conduct simulations. There is currently a lack of understanding of many environmental fate processes, such that it is sometimes difficult to construct appropriate models. For instance, indirect photochemistry and details of biodegradative processes are not well known. However, conceptually, it is generally possible to derive plausible models of environmental fate. The conclusion at this time is that there are a number of adequate environmental fate models that can be used directly or adapted for use with probability-based parameter estimates, but that the parameter estimates are often lacking. Thus, the greatest limitation for generating probabilistic distributions of EECs is generally more limited by input parameters than by the availability of models.

There are a number of errors of measurement and uncertainties in using simulation models to estimate environmental concentrations of pesticides. These range from systematic errors to random, unexplained variation. It is impossible to include all of the parameters and/or their possible nonlinear interactions, which may affect the concentration of the pesticide in the environment. It is, however, possible to construct canonical models which account for most of the important parameters that contribute the most to the variation in concentrations in the environment. There are literally thousands of parameters that can interact in a nonlinear manner, and they can vary spatially and temporally on scales of mm and seconds, respectively. These facts make it difficult to estimate environmental concentrations with a great deal of accuracy at the micro-scale, even if infinite resources were available to collect parametric information and even if a perfect algorithm were available.

3.4 STOCHASTIC MODELING

One modeling technique that is currently under development and that provides at least a partial solution to the problem of random error in the generation of EEC is **stochastic modeling**. This modeling approach recognizes that model inputs have natural environmental variation; therefore, this approach uses naturally occurring distributions of values as input instead of using single values to propagate the variability.

Propagated random errors in the modeling estimates of pesticide concentrations can be approximately quantified by

- first using sensitivity analyses to identify those input parameters for which input value changes most greatly change the output values and
- performing stochastic modeling by entering appropriate distributions rather than single values of important input variables (identified in the sensitivity analyses) into the model. The resulting output will also be a distribution that reflects the propagation of random errors in the sensitive input variables.

While the adoption of a system of frequency distributions to describe each parameter seems like an advancement and appropriate technique, several ramifications must be considered:

- Currently, there is a lack of information on a number of parameters necessary to conduct simulation such as for various fate parameters. There are many parameters for which accurate point estimates are not available. Thus, to advocate the use of stochastic estimates for all parameters is inappropriate. Instead, it is suggested that an analysis be conducted to determine which parameters influence the output to the greatest extent and are most amenable to the use of stochastic parameter inputs. Then, develop the required distributions for these parameters. Also, it should be remembered that there are several levels of variation among estimates of parameters. There are regional and seasonal variations that must be described. There are also variations that occur at the micro-scale and may represent measurement error.
- The results of stochastic modeling can be misleading if any significant correlation exists between any of the input variables and if the correlations are not accurately computed. If these correlations are not carried along, some of the computer runs in the random Monte Carlo simulation will have combinations of input values for different correlated input parameters that could not occur. Consequently, the resulting output from those runs would be meaningless.
- It must be remembered that the use of stochastic parameter estimates will have impacts on the statistics which are appropriate to be used in comparing the results of the modeling to the concentration-response relationships. Many environmental parameters are log-normally distributed. The additive and multiplicative interactions of stochastic parameters will produce results with probability distributions that are difficult to analyze. Thus, the adoption of the stochastic modeling process will require discussions of how to best interface the result of the fate models to the relevant toxicological information.
- One possible ramification of any move to the use of stochastic probability-based simulation modeling will be to develop estimates of EECs that have an associated variance described by a probability distribution. As the number of parameters with stochastic inputs increases, likewise the distribution range around the outputs will increase.

3.5 PROBABILISTIC EXPOSURE CHARACTERIZATIONS

As indicated in the introduction to this section, single values for EECs do not capture variability due to 1) changes in weather patterns at a given location or 2) changes in the characteristics of the agricultural site and body of water receiving runoff from that site. Unlike most of the random and systematic errors in computer estimates discussed in the previous section, this variation is completely natural and is reflected in vast changes in actual pesticide concentrations at a given location from year to year or across different locations in any given year.

To capture this variability, a probabilistic modeling approach should be used to develop EEC values that are defined in terms of likelihood of occurrence in the real world. There are essentially three methods of generating probabilistic exposure characterizations that reflect spatial and temporal variations in pesticide concentrations. One method is with the total use of stochastic modeling in which even large-scale spatial and temporal variations in sensitive input parameters over different sites/scenarios are described by distributions that are in turn input into the model instead of single values. However, as discussed in the previous section, there are a number of substantial problems associated with this approach. A second method to generate probabilistic exposure characterizations is to repeatedly run a deterministic model for single to multiple modeling sites/scenarios over multiple years. The third method of generating probabilistic exposure characterizations is a combination of the other two methods. It involves repeatedly running a stochastic model for single to multiple modeling sites/scenarios over multiple years. In this third method, only small-scale variations in sensitive input variables within each site/scenario are described by distributions. Variations across sites/scenarios are handled by separate computer runs, not by input distributions. At the present time, it is suggested that the second and third methods be used.

Repeated modeling over one to multiple modeling scenarios over multiple years provides a distribution of output values (*e.g.*, pesticide loadings from runoff/soil erosion models or EECs from surface water models) showing the frequency of their occurrence. A **modeling scenario** is defined as a specific combination of input values a model uses to produce one output or one distribution of outputs. The probabilistic modeling approach can utilize a single scenario that varies only in weather pattern so that EEC values reflect likelihood of occurrence at a specific site, or there can be different weather patterns and different soil and topographic combinations in order to express likelihood of EEC occurrence across a given region of the United States. Occurrence frequencies of interest may include the upper 50th, 25th, 10th, 5th, and 1st percentiles.

Another source of variation in EEC is the length of time over which an EEC is computed. This period can be hours or minutes for comparison to acute toxicological thresholds, or it can be several days or longer for comparison to chronic toxicological thresholds.

As previously discussed, the outputs of runoff/soil erosion and spray drift models include maximum storm event (or application loadings) for each year and annual pesticide loadings to surface waters. The outputs of surface water models can include maximum initial,

4-h, 12-h, 24-h, 48-h, 96-h, 21-d, 60-d, and 90-d average dissolved EECs for each year and site/scenario modeled.

If those models are run for one modeling site/scenario over multiple years, the resulting output distribution of pesticide loadings (from the runoff/soil erosion and spray drift models) or EECs (from the surface water model) can be presented as a cumulative frequency exceedance curve for each site/scenario and multiple years simulated. Each data point on a cumulative frequency exceedance curve for the simulation of one site/scenario over multiple years would have xy coordinates

y_i = annual maximum value of the output parameter at the site for any given year i and

x_i = fraction of years simulated with an annual maximum value of the output parameter $\geq y_i$.

A cumulative frequency exceedance curve for a single site/scenario over multiple years can be used to determine the probability that the annual maximum value of the output parameter will equal or exceed any given value of concern (*e.g.*, a toxicity threshold) at the site in any single future year and to determine the corresponding return frequency over multiple years.

For a single site/scenario modeled over multiple years, interesting values of the output parameter may include those with probabilities of being equaled or exceeded in any future year at the site of 50%, 25%, 10%, 5%, and 1%. Corresponding expected return frequencies over multiple future years include 1 in 2 years, 1 in 4 years, 1 in 10 years, 1 in 20 years, and 1 in 100 years, respectively. Within such a quantitative framework, such descriptive terms as "typical," "reasonable high-runoff scenario," and "highest runoff scenario" can be quantitatively defined as say the upper 50% (1 in 2 years), upper 10% (1 in 10 years), and upper 1% (1 in 100 years) exceedance probabilities (expected return frequencies), respectively.

Examples of cumulative frequency exceedance curves are given in Figures 3-3 and 3-4. In the following paragraphs, examples will be given on how each cumulative frequency exceedance curve can be used to determine the probability that the annual value of the output parameter will equal or exceed any given value of concern (*e.g.*, a toxicity threshold) at the site in any single future year and to determine the corresponding return frequency over multiple years.

In Figure 3-3, the total (dissolved and adsorbed) annual pesticide loss of the same pesticide from the same site for each of 36 years is plotted against the fraction of all 36 annual loadings equal to or exceeding it. Using the cumulative frequency exceedance curve in Figure 3-3, the probability that the total annual pesticide loss will equal or exceed any given value of concern at the site in any single future year can be determined, as well as the corresponding return frequency over multiple years. For example, it can be seen that there is approximately a 25% probability that the total annual pesticide maximum per-storm loss will equal or exceed 2.7 g/ha at the site in any single future year. That corresponds to an expected return frequency of 1 in 4 years over future years. Conversely, the loading corresponding to any percentile of interest can be determined. For example, the upper 10th percentile loading is approximately 4.8 g/ha.

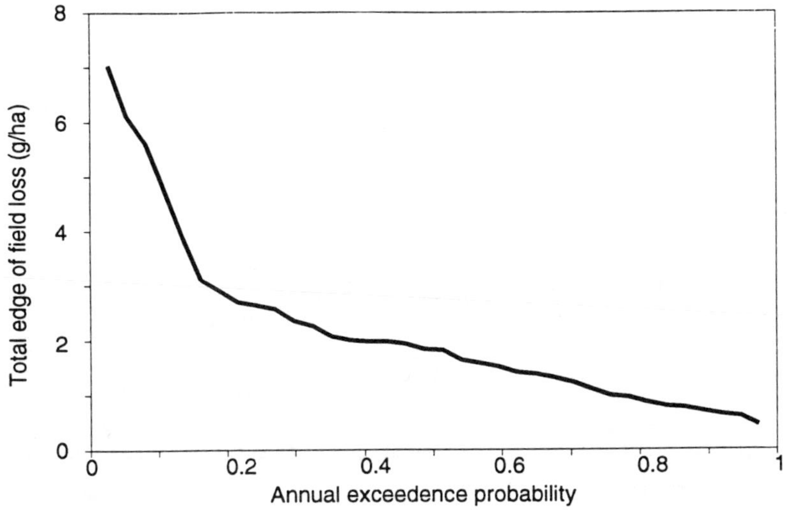

Figure 3-3. Annual pesticide loss (dissolved and adsorbed) in Mississippi (PRZM simulation on soybeans in Loring silt loam using a model pesticide)

In Figure 3-4, the 96-h average and 21-d average dissolved pesticide concentrations in a Mississippi pond receiving pesticide loadings from the Yazoo silt loam for each of 36 years are plotted against the fraction of all 36 maximum concentrations equal to or exceeding it. Using the cumulative frequency exceedance curve in Figure 3-4, the probability that the 96-h average and 21-d average dissolved concentration will equal or exceed any given value of concern in the pond (*e.g.,*, 96-h LC50 or 21-d MATC) in any single future year can be determined, as well as the corresponding return frequency over multiple years. For example, it can be seen that there is approximately a 24% probability that the annual maximum 96-h average dissolved pesticide concentration will equal or exceed the 96-h LC50 of 29 mg/L in the pond in any single future year. This probability corresponds to an expected return frequency of approximately 1 in 4 years over future years. It can also be seen that there is approximately a 23% probability that the annual maximum 21-d average concentration will exceed the 21-d MATC of 14 mg/L in the pond in any single future year. Conversely, the curves can be used to determine the maximum annual 96-h and 21-d concentrations corresponding to any percentile of interest. For example, the upper 10th percentile annual maximum 96-h and 21-d concentrations are approximately 41.5 mg/L and 19 mg/L respectively.

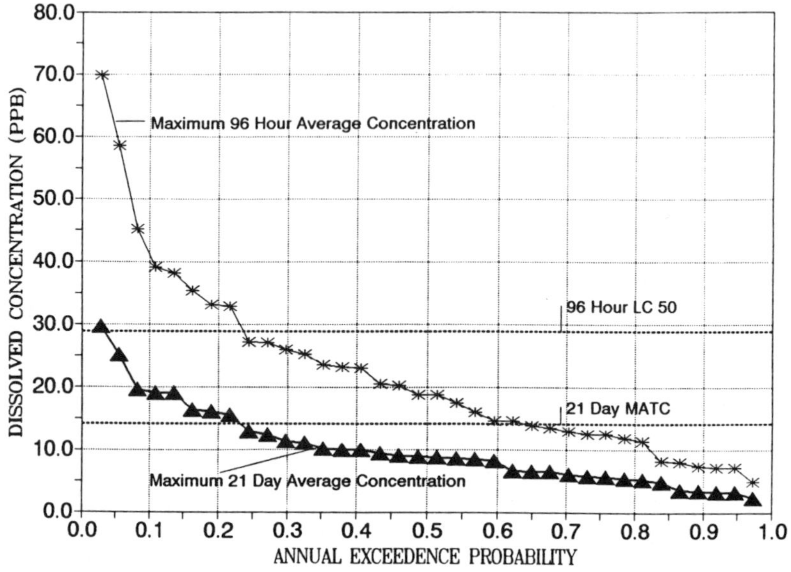

Figure 3-4. Maximum annual 96-hour and 21-day EECs versus fraction of years with maximum annual 96-h or 21-day EECs

If those models are run over multiple modeling sites/scenarios each over multiple years, the resulting output distribution can still be presented as a series of cumulative frequency exceedance curves, one for each site/scenario simulated over multiple years. The cumulative frequency exceedance curves for each site/scenario modeled over multiple years can (as previously described) be used to identify the upper 10th percentile EEC for each duration of interest (*e.g.*, 96-h, 21-d, etc.) and site/scenario modeled. The upper 10th percentile EECs can in turn be plotted as a cumulative frequency exceedance curve for each duration of interest over the multiple sites/scenarios modeled.

Each data point on the curve would have the following xy coordinates:

y_i = upper 10th percentile EEC for the duration of interest and site/scenario and

x_i = fraction of sites/scenarios modeled with an upper 10th percentile EEC > y_i.

The curve can then be used to identify those sites/scenarios with the largest upper 10th percentile EECs.

3.6 MODEL VALIDATION

The combination of systematic and propagated random error can be quantified by comparing the results of monitoring studies to those of computer modeling under the same conditions. Systematic and random errors can be caused by both field measurements and model parameterization (chemical- and field-specific parameters). If the differences are relatively small over a range of monitoring sites and corresponding modeling scenarios, the computer model is referred to as having been validated over that range. However, if substantial differences do occur, they may be due to field measurement errors and/or propagated uncertainties in the inputs (random errors) or to deficiencies in the model (systematic errors). If deficiencies in the model are a major source of error, differences between the results of the studies and computer modeling should generally be in the same direction (*e.g.*, generally positive or generally negative) indicative of systematic error.

At this time, validation of fate models suggests that they can be parameterized such that they can provide estimates that are consistently within one order of magnitude of the measured concentrations. At this time, it is uncertain whether this degree of resolution can be further improved or whether the greater source of the dissimilarity is arising from the field or modeling results. It is imperative that model validation continue in order to define more clearly whether fate models develop accurate EECs. For this, it is recommended that the generation of EECs via modeling be supported with a field experimentation program aimed at model validation.

Small-scale runoff studies are becoming increasingly popular because they are large enough to run comparative studies on the effects of formulation, tillage practices, crop cover, soil type, and slope on pesticide losses due to dissolved runoff and soil erosion, but small enough to allow for the use of weather-independent rainfall simulators. Small-scale runoff studies generally overestimate pesticide losses from actual agricultural fields, partly because their small size does not allow for as much in-field deposition of eroding sediment and the small-pocket ponding and infiltration of runoff water that occurs in full-size fields. However, recent research on the scaling effects of runoff studies suggests that much of the difference between results of small- and field-scale studies may be due to differences in experimental conditions (such as the use of intense artificial rainfall) rather than hydrological scaling effects. Such research also appears to be leading to accurate hydrological methods of extrapolating the results of small-scale runoff studies to field scale. Therefore, small-scale runoff studies probably can be used for field-scale model validation as well as for comparative studies and model calibration (Wauchope and Burgoa, 1992; Wauchope *et al.*, 1993; Sumner *et al.*, 1994).

Properly conducted field-scale runoff studies such as those performed using the methodologies discussed in the USEPA Field Agricultural Monitoring (FARM) Manual (Smith *et al.*, 1985) can be used to validate runoff/soil erosion models. Unfortunately, studies that monitor only farm ponds and not the surrounding watershed qualify only partially as field-scale runoff studies suitable for model validation. Studies that monitor only farm ponds involve monitoring of the pond water and sediment after storm-event pesticide loadings from the adjacent pesticide-treated fields, but they generally do not include the determination of runoff volume, sediment yield, and associated pesticide loadings from the field.

Surface water monitoring studies alone can sometimes be used to at least partially validate surface water models. However, if significant differences occur between the modeling and monitoring results, it is not possible to determine the relative contributions of the runoff and surface water models to the resulting overall systematic error, unless a runoff study of the same scale (field or basin) is also run in conjunction with the surface water monitoring study. Similarly, since numerous factors affect the concentrations of pesticides in surface waters, it is difficult to assess the effectiveness of mitigation methods through comparative surface water monitoring studies alone without also running runoff studies that use control (no mitigation) fields/basins as well as mitigated fields/basins. A good discussion on the relative merits and requirements of small-scale runoff, field-scale runoff, and surface water monitoring studies is contained in the final report of the Aquatic Effects Dialogue Group (Hendley, 1992).

Continued discussion is needed on how to validate models and on what constitutes sufficient validation. In most cases, the hydrology/soil erosion portions of the main line models that are typically used to simulate pesticide fate and transport have been, or soon will be, well validated. In many cases, only the pesticide fate algorithms need to be further validated. As previously discussed, that can be difficult because major discrepancies between modeling and field results may often reflect propagated uncertainties in the input values rather than deficiencies in the computer model. To minimize this source of error and enhance the chances for validation, it is suggested that key model input parameters such as K_d and soil degradation constant be characterized specifically for the soils at the field sites selected for validation.

If field and modeling results are generally close, a decision has to be made on what constitutes sufficient validation. For example, one might specify that differences between field observations and model predictions must be within a factor of two at least 95% of the time. Obviously, validation would be desired for pesticides with a wide range of fate characteristics (low to high persistence, low to high physical mobility, neutral to ionic, etc.) for several points throughout the distribution of estimated pesticide loadings and/or EECs. However, validation should not be necessary for every pesticide modeled.

The FIFRA EMWG is studying the current verification status of its recommended models and is evaluating what constitutes adequate verification for a given model with respect to the range of environmental conditions and pesticide properties it is expected to simulate. The EMWG will make recommendations to the scientific staff of OPP on the number, location, and types of additional studies required to adequately verify the recommended models. The EMWG will identify funding and personnel resources for conducting the recommended studies. The group will also approve study protocols, review study results, and prepare a final report on model verification.

3.7 POTENTIAL USES OF MODELING

Computer models for exposure of aquatic environments to pesticides are tools to generate estimates of EECs that can be used in a variety of ways. Within the tier framework described earlier, modeling procedures of varying complexity can be applied to risk characterization, depending on the need for greater realism and the possible consequences of the exposure. The simplest process may provide a single point estimate that can be part of a screening test or may be used to focus attention on the principal dissipation mechanism. More complex processes using a probabilistic approach serve to place EECs into perspective by providing a range of values and the likelihood that the EEC will be exceeded. In all cases, we are proposing the use of models to obtain a distribution of estimated environmental concentrations.

Models for generating EECs codify our understanding of transport and degradation and provide a link between the limited number of field observations that we can obtain and the wide range of possible conditions that may exist over the time period a pesticide may be used. Exposure estimates for a given pesticide can be obtained under a variety of conditions to characterize exposure under typical or extreme conditions. The effects of degradation rate and partitioning can be assessed as a function of weather and soil type. Models allow us to determine whether our understanding of dissipation mechanisms reflects reality. Are all modes of dissipation represented? Are lab rates of degradation providing a range that allows adequate predictions of field observations? What is the likely range of degradation rates? Studying pesticide behavior at a single site will be predominantly a study of mechanisms of transport (by water, soil, or spray drift), while consideration of multiple sites brings different soil types and geographical factors.

As more realistic estimates of exposure are required to justify the use of a pesticide, the use of models becomes increasingly complex. Once the dissipation and transport mechanisms have been described adequately, the exposure characterization is dependent upon detailed pesticide use data, geographical data (such as proximity of fields to water bodies), characteristics of receiving water bodies, crop plantings, tillage practices, agricultural management practices, and other land use information. Acknowledgment is made that the amount of information necessary to conduct a complete exposure characterization for a pesticide used on a major crop is difficult, if not impossible, to obtain.

The Aquatic Dialogue Group recommends the use of models in conjunction with field studies to increase confidence in the model predictions and, hence, increase confidence in the understanding of environmental fate and exposure characterization. Models may be used to compare pesticides with different properties applied to the same location under the same conditions. The power of the model, in this case, lies in its ability to recreate identical conditions under which the experiment will be run. The behavior of different pesticides under the same conditions can be compared, and expected behavior (or differences in behavior) can be predicted. The comparisons may be particularly powerful when field observations are available for one of the pesticides, in effect allowing the predictions to be validated by a field study. The known behavior of one chemical can be used to predict the environmental fate of another with greater confidence.

Modeling results under a standard set of conditions have been used to determine whether there are significant differences in the environmental fate of alternative pesticides. Given sufficient confidence in the understanding of transport and degradation mechanisms, field studies should not be necessary to establish similar behavior or detect significant differences in exposure. Comparison of different pesticides may have the greatest value in determining appropriate mitigation measures.

The proposed behavior scheme outlined in Chapter 2 illustrates another means of using models independently of field studies. The lower tiers should be standard modeling scenarios that can be obtained and interpreted easily. By using probabilistic modeling, the standard scenario can be set to represent typical cases or extreme cases with agreed-upon exceedance probabilities. The single point estimates of the lower tiers would be placed in context, and the overall risk characterization would be argued on probable effects.

Estimation of surface water concentrations, in principle, can be obtained similarly from a knowledge of runoff and spray drift, the principal input sources of pesticides to surface water. Since models for both input factors are based on empirical relationships with large uncertainties, modeling results have been viewed with skepticism, and validation of model predictions continues to be desirable. The degradation of the pesticide also can be modeled using known mechanisms. If model predictions are not accurate, then the model does not reflect reality, and additional information on pesticide degradation must be sought.

The development of field study protocols can be aided by models that predict lifetimes, concentrations, and possible variability. Models may play a role in selection of sites for field and/or monitoring studies. In the design of the study, sample times, sample numbers, and sample types must be set to accomplish the study's goal. Analytical methods must be developed with sufficient precision and sensitivity. Modeling results can be used to establish necessary detection and quantitation limits that will influence the development of the method. The interpretation of the results will depend also on whether the study has sufficient samples to give the desired level of precision. Model predictions can also be used to determine the types of samples necessary to characterize the variability in individual measurements and requirements for composite samples.

The current emphasis on mitigation to reduce exposure places new demands on modeling. Changes in agricultural practices can be examined to determine whether proposed mitigation will reduce exposure to acceptable levels. Ideally, mitigation measures could be applied in a computer model to determine the effect on exposure. If mitigation is effective in reducing exposure, the pesticide would be regulated accordingly. Verification of the effectiveness of the mitigation may be required, depending upon the level of concern and the success of the model in predicting the effects of a given mitigation measure.

Within the context of ecological risk characterization, modeling can be used to generate estimates of EECs which can be compared to toxicological endpoints. Caution must be exercised (see limitations in Section 3.8) in the development of low- to mid-tier aquatic risk characterizations through comparisons of estimated exposures to aquatic toxicity thresholds.

The Aquatic Dialogue Group recommends the use of Geographical Information Systems (GIS) to help identify geographic and landscape factors that affect pesticide concentrations in surface waters.

3.8 LIMITATIONS OF COMPUTER MODELING

Care must be taken to consider known and/or suspected errors and uncertainties as well as the natural variation of both the estimated exposures and toxicity thresholds if they are compared as part of an aquatic risk characterization; in using models to simulate extreme conditions outside the range of conditions from which the model was developed; and in using fate input values determined under a given set of experimental values to model a substantially different set of environmental conditions. Models must capture more effectively the influence of climate and soil on the environmental chemistry of pesticides.

Most of the computer models cannot adequately simulate (or simulate at all) complex hydrology, tile drainage, non-first order biodegradation kinetics, buffer zones, different formulations, some management practices, plant growth, or adsorption/desorption kinetics. Some of the input information, such as foliar dissipation, foliar washoff, and aquatic dissipation rates may not always be available.

It should be noted that even though the simulation of multiple sites/scenarios to give distributions of EECs instead of single EECs represents a major improvement in the aquatic risk characterization process, most of the estimated EECs within the distribution will probably be conservative since all assume that the surface water is adjacent to the field and that 100% of the drainage area is treated with pesticide.

Modeling of large watersheds is not well developed. If enough information is known about land use and percentage area treated in the drainage basin, more realistic basin-scale modeling using perhaps SWRRB could possibly be used in the higher tiers of the aquatic risk characterization. However, the only way to actually determine pesticide concentrations to which aquatic organisms may be exposed is through the use of well-designed monitoring studies in surface waters known to drain areas with a high percentage of pesticide treatment.

The new paradigm of aquatic risk characterization places the primary responsibility for generating EECs for comparison to aquatic concentrations levels on computer modeling. However, the greatest confidence in computer modeling currently remains in the comparison of EECs for different pesticides under the same conditions or for the same pesticide under different conditions. The creation of greater confidence in the use of model-generated EECs as absolute numbers (*e.g.*, as in their comparison to aquatic effects levels) will require substantially greater model validation. Field studies will be an essential component of the model validation process.

3.9 RECOMMENDATIONS

Procedures for estimating and using probabilistic EECs will benefit considerably from improvements in the use of EECs in risk characterization; improvements in models describing runoff, spray drift, and fate in surface water; and development of better data bases supplying input to the models. Recommendations to the developers and users of EECs are the following:

- Develop and use EECs on a probabilistic basis. Place future and existing modeling scenarios currently used for making risk decisions into a probabilistic framework. EECs developed from single scenario modeling procedures can and should be placed into perspective probabilistically. For example, "typical," "reasonable high-exposure scenario," and "highest exposure scenario" should all be defined quantitatively in terms of exceedance probabilities such as upper 50th, 10th, and 1st percentile, respectively.
- Provide adequate support and resources to develop and maintain modeling expertise within USEPA. Modeling is central to efficient and effective environmental risk assessment, and it must be resourced adequately within OPP for implementation and the realization of its benefits.
- Aim laboratory and field testing programs more specifically at the support of probability modeling and the validation of modeling results for runoff and fate in surface waters. Review and modify Sub-part N testing guidelines with these goals in mind, keeping in clear focus the need to develop information that contributes to EECs. Of particular interest are the areas of pesticide soil degradation and soil sorption/desorption since these two processes are very important parameters that impact pesticide movement off-site.
- Shift the emphasis of field environmental studies to validation of modeling results. Validate via an agrochemical industry community effort for data collection with USEPA and/or a third party serving as the focal point for validation. Model validation requires a considerable amount of suitable field data, and the most efficient way of achieving this is by pooling information collected by individual companies. Much effort has been expended in developing, parameterizing, and verifying environmental fate models during the last 20 years, but there is a great deal more research and verification needed.
- Develop readily accessible data bases and modeling systems that support exposure characterization on a national basis. The Aquatic Dialogue Group is aware that modeling systems combining GIS with environmental fate simulation models are under development. It endorses this activity and encourages its adoption into the risk characterization procedures. The Aquatic Dialogue Group also is aware of USEPA's efforts to develop better and more comprehensive data storage and retrieval systems, which should facilitate the development of stochastic estimates of parameters on a regional and national basis. The Aquatic Dialogue Group endorses the continued development of this type of data storage and retrieval system and emphasizes the need for quality control of the information stored in such a system. It is recommended that a liaison be established so that the data storage and retrieval system is compatible with that needed to conduct environmental fate modeling with a minimum of manual manipulation and interpretation. The continued development of these technologies and their attendant data bases will allow more sophisticated probability estimates of exposure in the future.

- Continue to improve environmental fate model algorithms. Current models have the following characteristics:
  - They have different levels of capability to predict mitigation effects from formulation changes, management practices, and buffer strips.
  - They do not handle plant effects such as pesticide washoff very well, nor do they deal with climatic effects on the rate at which pesticides degrade in soil.
  - Most cannot readily utilize the statistical distributions of parameters required for probability modeling.

The Aquatic Dialogue Group endorses the continued development of algorithms that are more effective in handling these issues, in addition to providing greater capacity to describe each parameter as a frequency distribution or discrete random variable.

REFERENCES

Hendley, P. 1992. Field studies used in calibrating models and calculating estimated environmental concentrations. In Improving Aquatic Risk Assessment under FIFRA, report of the Aquatic Effects Dialogue Group. World Wildlife Fund.

Smith, C.N., D.S. Brown, J.D. Dean, R.S. Parrish, R.F. Carsel and A.S. Donigan, Jr. 1985. *Field Agricultural Runoff Monitoring (FARM) Manual.* EPA/600/3-85/043. Prepared for and in conjunction with the USEPA Environmental Research Laboratory, Athens, GA by Anderson-Nichols, Inc., Palo Alto, CA, and Computer Sciences Corporation, Athens, GA.

Sumner, H.R., R.D. Wauchope, C.C. Truman, C.C. Dowler and J.E. Hook. 1994. A rainfall simulator and plot design for mesoplot runoff studies. Submitted to *Trans. Amer. Soc. Ag. Eng.*

Wauchope, D. 1992. Models currently available for developing estimated environmental concentrations for surface water runoff. In Improving Aquatic Risk Assessment under FIFRA, report of the Aquatic Effects Dialogue Group. World Wildlife Fund.

Wauchope, R.D. and B. Burgoa. 1992. Pesticide runoff studies: Toward a new protocol. Paper presented at ACS Meeting, Washington, DC, August 1992. Will be published in ACS Symposium Ser., American Chemical Society, Washington, DC.

Wauchope, R.D., C.C. Dowler, H.R. Sumner, C. Truman, A.W. Johnson and L.D. Chandler. 1993. Herbicide runoff measures from very small plots: how realistic? *Proceedings,* Brighton Crop Prot. Conference, Weeds, UK, 1993, pp. 1291-1298.

SELECTED REFERENCES

USDA-.SCS. 1992. Model evaluation report: Erosion/Productivity Impact Calculator (EPIC). Washington, DC.

USDA-.SCS. 1992. Model evaluation report: Simulator of Water Resources in Rural Basins - Water Quality Model (SWRRBWQ). Washington, DC.

USDA-.SCS. 1992. Groundwater Loading Effects of Agricultural Management Systems (GLEAMS). Washington, DC.

Chapter 4: MITIGATION OPTIONS

The purpose of this chapter is to (1) discuss some of the known mitigation options that, based on best professional judgment, will reduce the movement of pesticides off the site of application and therefore result in less risk to aquatic organisms (with some options only relevant to agricultural uses) and (2) address how verification of the effectiveness of these mitigation steps can be accomplished.

The basis for the **new paradigm** is immediate mitigation against identified, unacceptable risks—both potential and existing. Implementing steps to reduce or eliminate the identified risks is expected to protect the environment and save resources that in the past were utilized to demonstrate that the risk did not exist (which seldom was accomplished to everyone's satisfaction). Depending on the types, nature, and understanding of the identified risks and accepted mitigation actions, additional verification data may be required to support or rebut the U.S. Environmental Protection Agency's (USEPA) decision.

Risk reduction or **risk mitigation activities** are defined as actions taken to reduce or eliminate pesticide concentrations in aquatic and terrestrial habitats. Risk mitigation can involve numerous regulatory activities, from labeling restrictions to cancellation of specific uses. Mitigation measures should reflect the best practical nonpoint control technologies, processes, or other measures that will achieve the greatest reduction in the exposure to pesticides of nontarget aquatic organisms and habitats.

Recognizing the complex nature of mitigating aquatic risks and the limited number of informational sources regarding mitigation practices, this chapter offers an overview of mitigation options for reducing the probability of exceeding aquatic ecological levels of concern (LOCs). USEPA is currently implementing aquatic risk management policy in reaching registration and reregistration decisions.

4.1 GENERAL CONSIDERATIONS

Regardless of the type of mitigation proposed for a specific pesticidal use, the mitigation measure should exhibit several practical attributes. Mitigation goals should incorporate the concept of timely, value-added decisions with cost-effective utilization of financial and human resources. In addition, mitigation measures should be consistent with other federal, regional, state, and local resource protection and pollution prevention initiatives.

Mitigation should provide meaningful pollution prevention and a net benefit for aquatic resources. The process used to propose and implement mitigation should be clearly defined. Mitigation measures should be pragmatic and achievable. In addition, for any mitigation measures that require specific labeling language, the enforceability of such measures should be considered.

Practicality on the part of the regulator, the registrant, and the user must be considered when evaluating the need for and potential effectiveness of mitigation. Requiring mitigation based on preliminary or faulty risk characterizations can create undue burden, and thus cost, for the user. Mitigation for the sake of mitigation is not sound risk management policy. Often, the decisions of the regulator to require mitigation and the user to implement mitigation require the use of "best professional judgment." For example, in the area of erosion control, a possible mitigation step would be to require the use of the most appropriate best management practices (BMPs) for the region. Flexibility would be required, because the user and the local Soil Conservation Service (SCS) office should be relied upon for the most appropriate BMP strategy. Such decisions will likely be only qualitative and "directionally correct."

In order to propose effective mitigation measures, ecological risk assessments of pesticides must address several critical elements. Many of the facts collated during the characterization phase of risk assessment need to be evaluated when considering risk mitigation options. From the risk characterization, the following information should be available:

- types of impact expected and the aquatic organisms at risk;
- types and locations of aquatic habitat that may be adversely affected;
- primary routes of exposure (*i.e.*, drift versus runoff);
- use patterns, regions of the country where the use occurs, and typical pesticide application practices;
- potential influences of cultural practices, application procedures, and regional or local conditions on the routes of exposure; and
- relative reductions in exposure required to mitigate adverse impact to the receptors at risk.

This information should be used to identify the types of mitigation measures that may be effective in reducing exposure and thus risk.

When appropriate, the cluster concept should be applied to mitigation and to evaluating the effectiveness of mitigation. Clustering can be conducted based on use patterns and/or specific pesticide properties. An example of such clustering is the ongoing evaluation of avian hazard associated with granular insecticides used on turf and corn. Similarly, the USEPA is coordinating risk mitigation measures for all pyrethroids used on cotton. Such an approach helps to minimize resource expenditures of both the USEPA and the registrants, and it can also minimize confusion in the agricultural community regarding the specific mitigation practices required. There are numerous mechanisms for reducing aquatic nontarget exposure, and if each registrant takes a different approach, already complicated labels could become even more difficult for users to interpret and follow.

Limiting transport or delivery of a given pesticide from the edge of a field or a treated area to the water resource of concern may require the construction or establishment of terraces, detention ponds, manmade wetlands, buffer zones, vegetative filter strips, or grassed waterways. It is important to realize that these "zones" would be maintained for the duration of a particular use of the pesticide. If the label restriction were dropped or the pesticide use discontinued or the land use changed such that the restriction no longer applied, then it would be the owner's option to abandon the zone or return the land to its prior state.

The first and foremost consideration for mitigation is actually reducing the amount of product applied in a single application or season. Once the label is optimized, mitigation should attempt to minimize off-site movement. The primary routes of off-site movement are spray drift and surface runoff. Numerous mitigation options are available for reducing exposure from these different sources.

4.2 GENERIC MITIGATION MEASURES

A summary of generic mitigation measures can be found in Table 4-1.

TABLE 4-1. Summary of Generic Mitigation Measures for Reducing Off-site Movement

Mitigation Factor	Rationale or Comments
Reduction in Application Rate	Reduce total pesticide load to the environment; reduction in exposure proportional to rate reduction.
Reduction in Number of Applications	Reduce total pesticide load to the environment; maximum exposure levels may be similar, but probability of occurrence will be less and chronic exposure may be reduced.
Integrated Pest Management	Reduce total pesticide load to the environment.
Increased Application Interval	May decrease chronic exposure.
Regional/Local Restrictions	Restrict use of product in highly sensitive areas.
Crop-Specific Restrictions	Restrict use on crops in sensitive areas and/or on crops that require numerous applications or higher rates.
Prescription Use	Require permission by state or local officials prior to use; minimize use in sensitive areas or when conditions favor off - site movement.
Education/Training Programs	Minimize misuse via increased training/education; expedite implementation of mitigation options.
Equipment Calibration	Reduce over-application and off-site movement by ensuring equipment is properly calibrated.

4.2.1 Chemical Use

There are four major spray application scenarios:

1. ground spray boom,
2. chemigation,
3. airblast, and
4. aerial.

With these methods of application, movement of pesticide off-site may be by drift or runoff or both.

A common approach for reducing levels of concern for agrochemicals is **source control**: reducing the rate and frequency of pesticide application. Secondary to direct source reduction are measures taken to reduce the movement of pesticides to surface water (field and delivery control) or to make pesticides less available to aquatic habitats where they affect nontarget organisms. The two primary methods of source control are

1. reduction in the application rate and
2. reduction in the number of applications.

There are cases in which risk mitigation has been approached primarily by reducing application rates. In general, reduced application rates have been and will remain the starting points for risk mitigation. Such reductions may require a commitment by the registrant to develop efficacy data supporting a wider range of use rates. This effort to "optimize" use rates, enabling users to fine tune their applications, represents the kind of commitment to long-term source reduction that should be promoted. It is also possible to provide new instructions that include a range of applications for control. For example, there could be recommendations for "split" applications, where a second application is recommended only if the pest reoccurs.

Reducing the number of allowable applications has also been used as a risk reduction strategy. Reducing the number of applications decreases the total environmental load and may decrease potential acute and chronic risk. Without changing the rate per application, peak acute exposure values may or may not be reduced when the number of applications decreases; however, the probability of runoff events and the number of potential spray drift events will certainly be reduced. With such a decrease in the number of exposure events, recovery from acute exposure may be greater, and chronic exposure may be reduced.

Depending on the specific pesticide characteristics and the type of toxicity of concern, a third mitigation measure in this category might be

3. increasing the application interval.

Increasing the application interval may decrease the accumulation of moderately persistent pesticides, but in some cases may result in the exposure of more animals for a longer period of time. Therefore, the true risk reduction associated with this mitigation option needs to be carefully evaluated.

Other source reduction mitigation measures, which may be useful in cases where it is not possible to reduce application rates for specific uses, include the following:

- Restrict pest spectrum such that maximum acceptable rate can be reduced. Such restrictions will generally result in decreased use of the product, possibly in more "susceptible" or sensitive areas, or a decrease in the application rate required for the control of the more difficult pests.
- Limit the total number of acres treated within a watershed. This approach will result in less load on the environment, although it can be difficult to enforce/regulate such a restriction. For example, restricting total poundage used within a watershed may require some type of prescriptive use.
- Impose manufacturing limitations or caps. This is obviously a direct control of the amount of product used, although it may not directly reduce the potential exposure in a given area.

- Implement state management plans. This approach, currently being used primarily for ground water protection (although surface water quality concerns are also to be addressed), provides the state with control over the specific mitigation plans that may be necessary to achieve reduced exposure for their specific circumstances.

Mitigation measures in each of these categories may reduce overall usage, even if specific application rates are unchanged.

To support source reduction mitigation measures, data on acreage treated and evidence that a likely result of the mitigation measure is a reduction in number of acres treated are necessary. Such data, including information on maximum versus typical use rates and estimates of acreage affected by the proposed label changes, may need to be developed.

4.2.2 Regional or Local Restrictions

In certain situations, the risk characterization may indicate that the LOC is exceeded for a particular type of habitat or a particular watershed, or that an endangered species in a defined area might be threatened. In these cases, an acceptable level of risk may be achieved through limiting or altering the use of the pesticide. Examples of situations that may lend themselves to regional or local restrictions are tidal salt marshes or Midwestern potholes. Under these conditions, risk reductions may be affected by restrictions that limit the use of the pesticide in localities or by developing a unique use regime. In some cases it may be necessary to prohibit the use of the pesticide in certain regions of the country or counties within a state. These types of restrictions would require consultation between the registrant, the states involved, and the USEPA to ensure the practicality of such restrictions.

4.2.3 Crop-Specific Restrictions

The risk characterization might indicate that the LOC has been exceeded due to a particular use. That is, the concern results from the manner in which the pesticide is used and from the fact that a particular application is made in areas where endangered species or particularly sensitive ecosystems exist. In those cases, prohibition of use of the pesticide in certain areas or regions of the country may be adequate to reduce the LOC to acceptable risk. In extreme cases, cancellation of a particular use of a pesticide may be required to achieve an acceptable risk.

4.2.4 Prescription Use

If there are overwhelming benefits for retaining a particular pesticide use despite high ecological risk, prescription use may be a mitigation option. Due to the relatively high costs of operation and implementation, prescription use is only practical when the pesticide product is essential. In the case of prescription use, application would be allowed only under narrowly defined circumstances. Where prescription use is to be attempted, a recognized mechanism for the approval of the specified uses must be established. This mechanism must involve a state agency.

An example of a plan for prescription use is the use of azinphos-methyl on sugarcane in Louisiana. The opportunities for aquatic risk mitigation were limited, while the benefits, as judged by the state, were high. Therefore, one proposed risk mitigation strategy was to control total azinphos-methyl application in a given area by issuing permits for use. Mechanisms issuing permits for use, as well as tracking recommended and actual applications, had to be developed.

4.2.5 Educational and Training Programs

Educational and training programs are a useful aspect of any mitigation effort and can be used to inform pesticide applicators, growers, and the general public of

- the ecological risks associated with the use of the particular pesticide.
- the relevant factors that affect the ability of the user to mitigate these risks under very specific use scenarios. Programs can include education on the specific mitigation factors associated with a specific product or can generally address how factors such as proper equipment calibration can actually lead to less pesticide use and thus less drift.
- the components of integrated pest management (IPM) programs as related to a particular chemical, use, and pest. In particular, educational programs should focus on basing rates of treatment on thresholds as determined by scouting. If application threshold data are available, applications should be made when an economic threshold level has been reached, as opposed to applying pesticides in anticipation of pest problems. The aim should be to reach economically acceptable, rather than total, pest control.

Educational programs have been used as part of mitigation efforts that have been implemented for various products.

Educational programs are not a direct mitigation measure, but they have the advantage of often reducing other adverse environmental impacts of agriculture such as erosion and nutrient loading to surface water. The promotion of specific BMP's may be potentially useful in educational programs. Examples of potentially useful practices include soil conservation and related runoff control practices and integrated pest management as applied to specific chemicals, uses, and site conditions.

The most useful education programs incorporate pest management strategies that focus on diagnosing pest infestation levels to determine whether pesticide applications are needed. If pesticides are needed, only the amount of pesticide necessary should be applied. Educational material is most useful if widely distributed and incorporated into training programs for certified applicators.

4.2.6 Research and Compensatory Mitigation Options

There may be cases in which the registrants will find it useful to fund options that are not directly or immediately tied to risk mitigation. Any areas for funding should address long-term and/or complex research projects, such as the Spray Drift Task Force (SDTF). Funding of compensatory mitigation offers another potential area that addresses efforts to compensate for the ecological losses or expected losses associated with the use of a specific product or class of products.

There has been research into the effectiveness of several mitigation methods such as the reduction of nutrients, sediments, and other pollutants from surface waters. However, relatively little effort has been directed toward determining the effectiveness of mitigation methods in reducing pesticides concentrations. Field studies associated with risk reduction activities may help to answer the question of how effective such measures are for pesticide removal, as well as provide information on the optimal matrix of pesticide and mitigation measure characteristics. However, primary consideration should be given to funding existing research and practical state or federal monitoring programs that are well established, but which do not currently include monitoring for pesticides concentrations.

Modeling may be a useful research area to pursue in conjunction with monitoring. Models may be developed to predict effects of risk reduction measures under specific use conditions, to locate areas with the greatest potential problems, and to extend results beyond the small scale upon which monitoring is usually done.

Compensatory mitigation is a possibility for pesticides that exceed levels of concern but, due to important benefits, continue to be used even though there may be a lack of effective mitigation measures available. However, the necessity for compensatory mitigation arises largely as a risk management decision based on cost/benefit analysis. One way to approach compensatory mitigation is to value land, acre for comparable acre, in terms of its ecological value.

4.3 APPROACHES FOR MITIGATING RUNOFF

This section reviews the possible options for reducing the runoff loss of pesticides from agricultural fields. In recent years, other groups have also addressed the usefulness of the various approaches discussed below. Of particular relevance is the manual entitled "Guidance Specifying Management Measures for Sources of Nonpoint Pollution in Coastal Waters" (USEPA, 1992).

Many factors interact to determine the ultimate loss of pesticides from agricultural fields. Such factors and their interactions need to be understood to make appropriate decisions on potential mitigation options. These factors can be divided into three categories:

1. Pesticide properties—The major pesticide factors determining concentrations and losses of pesticides being transported with water from cropland are their persistence and adsorption to soil.
2. Hydrologic conditions—The major hydrologic factors are the intensity and duration of precipitation and the rate of infiltration as affected by soil factors and conditions.
3. Management practices—The major management factors to reduce field losses of pesticides are the rate, method, timing, and choice of applied pesticides; cropping sequence (in time and space); water management; and tillage or soil

management systems. The major management factors to reduce field-to-stream transport of pesticides are structures such as terraces and use of landscape reconfiguration such as wetlands or buffer strips.

To determine the best combination of mitigation practices to reduce pesticide losses, mechanisms of loss and pesticide interactions with the soil and mitigation practices must be understood. Depending on soil adsorption in relation to other factors, pesticides can be mostly lost with surface runoff water, sediment, or water percolating out of the root zone (water that may return to the surface through base flow or artificial subsurface drainage). Erosion control is a good mitigation practice for strongly adsorbed pesticides. For both strongly and moderately adsorbed pesticides, soil incorporation is a good mitigation practice that reduces the amount of pesticide in the thin surface soil "mixing zone" and reduces surface runoff losses. Other practices that enhance infiltration also reduce surface runoff losses. The route of infiltration (*e.g.*, through "macropores") can allow pesticides to percolate through the root zone more quickly than normally expected, resulting in "concentration spikes." However, for moderately to strongly adsorbed pesticides, leaching losses are usually much less than losses in surface runoff. For weakly or non-adsorbed pesticides, increased infiltration reduces runoff losses but may enhance leaching losses. If the macropores are earthworm channels, increased organic matter on the walls may adsorb some of the leaching pesticide.

4.3.1 Pesticide Factors

The two primary chemical factors affecting the fate of applied pesticides, which must be understood to determine the most appropriate mitigation option, are persistence and adsorption to soil. Persistence of a pesticide is determined by its properties, including how resistant it is to microbiological and/or chemical breakdown; its solubility, which may affect its availability to be broken down or lost with water; and its vapor pressure, which affects potential volatilization losses to the atmosphere. Breakdown or degradation of a pesticide to its degradation products or metabolites can be a concern if those metabolites also have toxic properties in the environment. Volatilization can be a particular concern when the pesticide is applied to plant residue rather than to the soil where it can be adsorbed. How quickly rainfall washes the pesticide from the crop residue to the soil will affect volatilization and persistence. Volatilized pesticides may be returned to earth with precipitation. As a first approximation, overall pesticide dissipation from the soil is assumed to be a first-order reaction, which means the rate of dissipation is proportional to the amount present. The proportionality constant is inversely related to "half-life," which is the time it takes for half the pesticide to dissipate.

The degree of soil adsorption is not only important in reducing possible pesticide volatilization, but it also largely determines how pesticides may be transported with excess water from treated fields. Based on their soil adsorption, pesticides can be grouped as

- strongly adsorbed—lost mostly with sediment;
- moderately adsorbed—lost mostly with surface runoff water; and
- weakly or non-adsorbed—lost mostly with percolating water.

As a first approximation, the concentration of soil-adsorbed pesticide is assumed to be directly proportional to its concentration in water under equilibrium conditions in a soil-water-chemical mixture. The **proportionality constant** (defined as K = the concentration in sediment or soil divided by the concentration in water) is often called the partition or adsorption coefficient. The properties of a pesticide primarily determine its potential to be adsorbed to soil. Soil properties (*e.g.*, organic matter content, pH, and clay content) and environmental conditions (*e.g.*, temperature and soil water content) exert a significant effect on both adsorption and pesticide persistence. With conservation tillage, surface crop residue can intercept sprayed pesticides, preventing them from reaching the soil and altering their fate.

4.3.2 Hydrologic Factors

The relative rates of precipitation and infiltration into the soil during a rain storm are the primary hydrologic factors affecting pesticide concentrations and losses by the modes shown in the simple schematic of a portion of the hydrologic cycle (Figure 4-1). A third factor would be the timing of the storm relative to pesticide application.

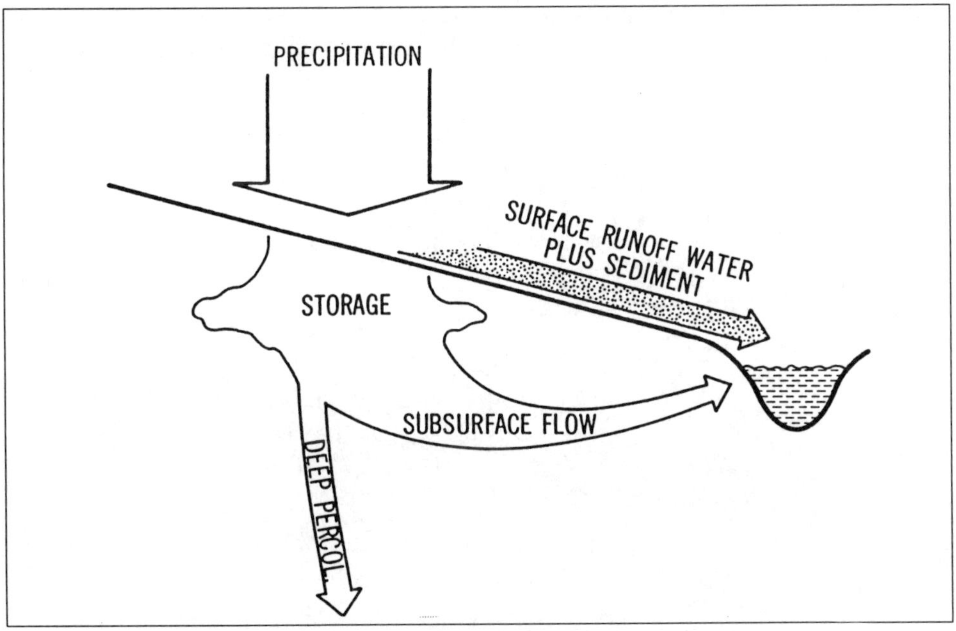

Figure 4-1. Simple schematic of a portion of the hydrologic cycle

The infiltration rate relative to rainfall intensity or rate determines the time of initiation and the quantity of runoff (and by subtraction of runoff plus soil water storage from precipitation, determines the quantity of excess water leaching through the root zone). Soil factors such as texture and soil conditions such as degree of compaction both affect infiltration rates (sandy, high-water-table soils may be a special case where all the rainfall infiltrates until their storage capacity is exceeded and runoff begins). Figure 4-2 helps illustrate the importance of the time runoff begins as well as the route of infiltration. A fairly thin layer of surface soil, called the "mixing zone," often assumed to be about 1 cm thick, interacts with and releases pesticide to rainfall and overland flow. For moderately and for particularly weakly or non-adsorbed pesticides, the amount of pesticide remaining in the mixing zone decreases with time and the amount of water moving through the zone during a storm. Hence, the pesticide concentration in initial runoff water decreases with delay in runoff; furthermore, the decrease in concentration with time during runoff, once it begins, is primarily a result of leaching of the pesticide from the mixing zone. Reduction in the amount of pesticide in the mixing zone between the time of application and the first storm (and between storms) through degradation, volatilization, and/or movement by diffusion explains why pesticide concentrations and losses (depending on runoff volumes) are generally the greatest in the first runoff event after application and decrease with time during the season. As shown in Figure 4-2, if the surface soil becomes saturated, some water may move through preferential flow paths or "macropores" and leach deeper or more quickly than would be expected if the water had to flow through the whole soil column, displacing the water below it. Conversely to increasing pesticide leaching, flow through macropores could reduce leaching if the pesticide is within the soil mass or aggregates and water moving through macropores bypasses it.

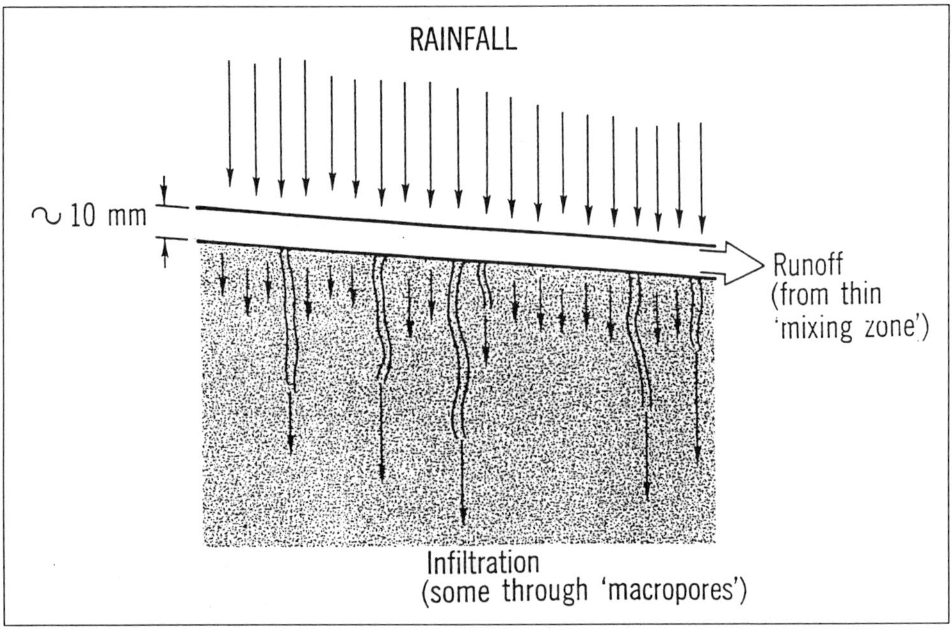

Figure 4-2. Schematic of soil surface showing mixing zone and macropores

4.3.3 Management Factors—Field Loss Reduction

Of all the management factors that affect a particular pesticide concentration and loss from the field, the rate of pesticide application is the one factor that has the most direct and usually the greatest effect. The method of pesticide application can also have a major effect on surface runoff if it affects placement relative to the route(s) of infiltration. Timing of pesticide application relative to subsequent storms is another major factor that can affect concentrations and losses. Where possible, choice of pesticide and/or formulation (or additive) used, by affecting pesticide properties, can affect concentrations and losses. Cropping sequence in both time and space (*e.g.*, contour strip cropping)—by affecting the needs for tillage, soil moisture, and weed and pest control—affects both hydrology and pesticide inputs and, thus, concentrations and losses. Tillage, by affecting erosion rates and hydrology as well as needs for and method of pesticide application, can potentially affect concentrations and losses. Water management in terms of drainage and irrigation also affects hydrology and thus potential losses.

In the following sections on mitigation practices to reduce pesticide runoff losses from treated fields and to reduce the field-to-stream transport of pesticides that are lost from treated fields, a rough estimate of the likely range of effect for each practice is given. It must be remembered this is just an estimate based on limited research (see References) and/or professional judgment. A better estimate can be made for more specific sets of conditions. Computer simulation models do exist that can predict pesticide fate and loss, with varying degrees of accuracy, for specific conditions and practices (see Chapter 3 for additional details).

4.3.3.1 Rate/Formulation

Lower rate feasibility. Rate of pesticide application is one of the primary factors affecting pesticide runoff losses. Pesticide runoff loss with sediment and water is roughly proportional to the amount of pesticide in the thin mixing zone at the soil surface and/or on crop residue or foliage. For example, if the amount applied is reduced by a factor of two, loss should be reduced by a factor of two. If this relationship deviates from linearity, losses should be reduced even more than the percentage the rate is reduced, because the degree of soil adsorption usually increases as concentration (or the amount applied) decreases. The feasibility of lower rates is dependent on both the amount needed to achieve the level of pest control desired and the accuracy of the application method. Realistically, a rate reduction beyond 25 to 50% is not likely, assuming the same pest control spectrum is maintained. A reduced application rate can reduce other environmental losses (*i.e.*, to groundwater and the atmosphere) and potential adverse biological effects, as well as reduce energy and economic inputs. However, reduced application rate can increase the risk of inadequate control or resistance and economic losses.

The factors of weather, hydrology, soil, and chemical properties should not greatly influence the predicted effect of rate reduction.

Partial substitution. Partial substitution of the pesticide of concern by another pesticide or by other means of pest control can reduce the amount applied. The effect is the same as above, where the predicted effect on pesticide runoff loss with sediment and water is roughly proportional to the amount of substitution and therefore the amount of pesticide in the thin mixing zone at the soil surface and/or on crop residue or foliage. If suitable alternatives exist, it is feasible to reduce the rate of applications beyond that of elimination alone, possibly up to 80%.

Again, a reduced application rate can reduce other environmental losses (*i.e.*, to groundwater and the atmosphere) and potential adverse biological effects as well as energy and economic inputs; there is also the potential for improved pest control with a mixture of pesticides or control methods. At the same time, there is the risk of reduced pest control with partial substitution; also, if one pesticide is substituted for another, the potential environmental impact of the substitute may be greater because of increased losses and/or greater toxicity.

If the combination of weather, hydrology, soil, and chemical factors favor greater loss of a substitute pesticide, the predicted benefit of this practice would be lessened.

Partial treatment. Partial treatment, either on a field basis or within a row (*e.g.*, band versus broadcast or directed spray versus an over-the-top application), can effectively reduce the rate of pesticide application, and again pesticide runoff loss with sediment and water is roughly proportional to the amount of pesticide in the thin mixing zone at the soil surface and/or on crop residue or foliage. Where a pesticide is applied in a band over only a portion of the row-width, realistically a 75% reduction is probably a maximum, compared to a broadcast application where the whole area is treated. "Spot treatment," seed treatment, or directed spray may allow greater reductions.

Again, a reduced application rate can reduce other environmental losses (*i.e.*, to groundwater and the atmosphere) and potentially adverse biological effects, as well as energy and economic inputs. However, a reduced application rate can increase the risk of pest escapes and economic losses, plus the untreated area generally requires alternative treatment (*e.g.*, timely mechanical cultivation for weed control), which may increase the potential for runoff and soil erosion.

The potential for alternative treatment will be influenced by weather, hydrology, and soil factors that affect trafficability in the case of ground-applied treatments.

Choice of formulation. If different formulations are available, a formulation that alters soil adsorption and solubility can affect pesticide runoff losses. The degree of reduction will depend on the major transport mechanism for a certain pesticide in relation to the differences in its adsorption and solubility properties for the different formulations. Encapsulation represents a type of formulation that can directly influence the availability of a product for either leaching or runoff. Runoff loss reduction up to 50% is probably achievable for some specific formulations, but measurements need to be made to verify their effect.

Other environmental losses (*e.g.*, volatilization) may also be reduced. However, it is also possible that other environmental losses may be increased, and pest control may be reduced.

Weather, hydrology, and soil factors are important. Where they combine for good erosion control, increased soil adsorption is a positive factor. Where these factors combine for good infiltration, increased solubility should reduce surface runoff losses; however, leaching losses may be increased.

***Soil erodibility special restrictions*.** Rate restrictions tied to highly erodible land unprotected by crop residue have the potential to reduce runoff losses. This targeting to areas where soil erosion is expected to be severe and where application rates may be restricted by some percentage (*e.g.*, 25%) should reduce pesticide runoff losses. In this example, the reduction would be at least 25% if maximum label rates were being used.

By targeting erodible areas that are likely more susceptible to both runoff and erosion, this restriction may be more beneficial than a blanket restriction in total if a negative effect in terms of greater risk from pests results from the reduced rate. Use of erosion control measures may also be promoted by this approach. However, a reduced application rate can increase the risk of pest escapes and economic losses. Frequently, the pesticide requirement for highly erodible soils may be greater than for other soils because pesticides are used for weed control in lieu of tillage. Pesticide selection becomes more critical for such soils.

The success of this targeting is related to weather, hydrology, and soil factors. A final important factor (*i.e.*, pesticide properties) is that soil adsorption determines the potential for transport with eroded soil and the feasibility for this approach to be effective.

4.3.3.2 Method and timing of application

***Soil incorporation*.** Mechanical incorporation (*e.g.*, pre-plant incorporated [PPI]) or incorporation with irrigation, by reducing the amount of pesticide in the thin mixing zone at the soil surface and/or on crop residue or foliage, reduces the interaction with and transfer of pesticide to runoff water and thereby reduces runoff loss. Reductions in the range of 25 to 70% are possible.

With incorporation, other environmental losses may also be reduced, and pest control may be improved. However, mechanical incorporation with tillage may reduce soil protecting crop residue with conservation tillage, and pest control may be reduced. Leaching losses may—but are not likely to—be increased, and the relative importance of the different loss pathways needs to be considered.

The feasibility of using incorporation as a mitigation practice depends on the pesticide mode of action relative to position in the soil profile, and the potential for incorporation with irrigation depends on sufficient infiltration and solubility (in conjunction with limited soil adsorption) of the pesticide. The use or availability of adequate irrigation can also be site specific.

***Timing relative to expected/pending storms or irrigation applications*.** For some situations and certain pesticides, a choice of timing of pesticide application(s) is possible, *e.g.*, pre-emergence versus post-emergence. Because the first storm after pesticide application will generally result in the highest pesticide concentrations in runoff and because concentrations will decrease with time both during and between runoff events, increasing the expected time interval between application and intense storms causing significant runoff could reduce pesticide runoff losses in part because intervening smaller, more gentle rains may move pesticide down out of the thin mixing zone at the soil surface. An additional positive effect is possible with post-emergence application of herbicides, where the weeds that are allowed to grow before treatment provide some cover for the soil to reduce runoff and erosion similar to crop residue. Reductions up to 50% may be possible.

Other advantages include possibly extending the time for other operations and possibly improved pest control. However, poor timing may also reduce pest control or increase other environmental losses.

The historical timing of runoff-producing storms along with the mode of action and persistence of the pesticide will determine the feasibility of this practice (dependent on weather, hydrology, soil, and pesticide factors).

4.3.3.3 Tillage systems

No-till. The use of no-till, where only minor soil disturbance occurs at planting and all the crop residue from the previous crop is left on the soil surface, is increasing. Erosion and pesticides transported with sediment are usually reduced 50 to 90% with no-till. Runoff volume, at least on a growing season or annual basis, is generally less than that from conventional moldboard plow tillage. While pesticide concentrations in no-till runoff may be higher, the total quantity of pesticide in runoff from no-till is usually less—sometimes much less—than from conventional tillage. The expected reduction in pesticide loss would be in the 0 to 90% range.

The other advantages, besides erosion control, include reduced time and energy inputs with tillage. However, use of no-till may

- require additional pesticide use (not an increase in rate of those used, but another product such as a "burn-down" herbicide);
- make it impossible to incorporate pesticides by conventional means, therefore eliminating possible use of some more "environmentally friendly" pesticides;
- under some conditions, increase runoff and volatilization of pesticides intercepted by surface crop residue;
- increase runoff volume(s) for first storm(s) after chemical application relative to soil loosened by tillage; and/or
- result in more macropores at the soil surface, which may enhance deeper leaching of pesticides (*i.e.*, pesticides are detected more quickly and at deeper depths than expected based on their adsorption to soil).

However, if no-till results in use of a different "mix" of pesticides (lower-rate, post-emergence herbicides, for example), the total pesticide runoff loss may be decreased depending on individual pesticide properties. In addition, the strong soil adsorption properties of the "burndown" herbicides most often used, paraquat and glyphosate, result in little environmental concern for runoff losses with the no-till system.

When soil (including slope) and weather conditions are conducive to erosion, and for strongly adsorbed pesticides, no-till has its greatest potential for reducing pesticide runoff.

Conservation tillage. Conservation tillage (*i.e.*, reduced tillage or mulch tillage) involves some tillage but less than that of inversion moldboard plow tillage. To qualify as conservation tillage, at least 30% of the soil surface must be covered with crop residue after planting. The reduction in runoff and erosion expected with conservation tillage reduces the carriers and therefore runoff losses of pesticides. The likely range of reduction in runoff is 0 to 75%. Ridge tillage may be a special case where, in addition to reduced pesticide losses from reduced runoff and erosion, reduced herbicide use through banding (feasible and generally done because of later ridge rebuilding cultivation) further reduces losses.

Advantages besides erosion control include reduced time and energy inputs with tillage. However, as with no-till, conservation tillage may require additional pesticide use (possibly, a "burn-down" herbicide); reduce the ability to incorporate pesticides by conventional means, possibly eliminating the use of more "environmentally friendly" pesticides; under some conditions, increase runoff and some volatilization of pesticides intercepted by surface residue; and/or result in more macropores at the soil surface possibly enhancing deeper leaching of pesticides, but to a much lesser degree than no-till.

This practice has the most potential for reducing runoff when soil (including slope) and weather conditions are conducive to erosion and when strongly adsorbed pesticides are used.

4.3.3.4 Water management

Subsurface drainage. Artificial subsurface drainage, achieved mainly by removing water from the soil profile between rainfall events by using drainage tubes and ditches, lowers the soil antecedent moisture content and somewhat increases infiltration during a rainfall event. Therefore, subsurface drainage delays and reduces the volume of surface runoff and forces water to move through the soil profile, all of which generally reduce both pesticide concentrations and losses. In the case of weakly absorbed pesticides and sandy, low organic matter soils, concentrations may not be decreased. Losses might be reduced from 0 to 50% with subsurface drainage in areas where subsurface drainage is useful.

Other advantages are that subsurface drainage makes possible more timely field operations and provides for better soil moisture conditions for crop growth. Drainage is needed and/or used only in areas with potential for water-logged soils; the beneficial effect of subsurface drainage in reducing pesticide runoff losses is generally greater for moderately adsorbed pesticides.

Surface sealing/compaction. Surface sealing of unprotected soil and compaction due to traffic can reduce infiltration rates. Thus, surface sealing and/or compaction decreases the interval between initiation of rainfall and initiation of runoff and also increases the rate, and therefore the volume, of runoff. Since pesticide concentrations decrease with time during a runoff event (as pesticide is leached out of the thin mixing zone at the soil surface), more runoff, particularly early in the rainfall event when concentrations are higher, results in greater pesticide loss. Protecting the soil surface with living or dead crop residue absorbs rainfall energy and reduces sealing. Reduced or controlled traffic can reduce compaction.

Other advantages of controlling surface sealing/compaction include more infiltration and better soil aeration, which generally mean better growing conditions for crops and likely more microbiological activity. There are usually no disadvantages.

The problem of surface sealing/compaction is more important on soils with good internal drainage in areas with excess precipitation. Potential problems associated with sealing are worse for less strongly adsorbed pesticides.

Irrigation. Irrigation management practices can have a significant effect on the quantity and quality of water leaving the field associated with runoff. Specific practices that should be considered include the following:

- Surge flow—as an application method of pulses used in furrow irrigation. The goal is to increase application uniformity along the furrow and reduce the amount of induced runoff, erosion, and percolation (and therefore, both runoff and leaching pesticide losses; this practice usually reduces total water input).
- Low-energy precision application (LEPA)—as a type of center pivot or linear irrigation system, which increases application efficiency and reduces the amount of excess water that is applied. This practice will minimize the potential for runoff.
- Drip or microjet—or other precision methods for irrigating that reduce the amount of excess water applied, thus reducing runoff, erosion, and percolation losses
- Furrow diking—the practice of mechanically making small dikes within the furrows to pond water, thus eliminating (in some areas it is illegal for irrigation tailwaters to leave the site) or reducing runoff and erosion
- High-volume traveling or overhead guns—apply large volumes of water in short time. Where runoff is likely, more frequent applications of shorter duration will minimize the potential for runoff.
- Retention ponds—the practice of collecting irrigation tailwaters for reuse to eliminate runoff leaving the field
- Chemigation—On very permeable or very impermeable soils with high slope, the probability of leaching or runoff, respectively, is high. If this method of application is considered for registration, physical properties of the product and soil type will determine the mitigation options.

4.3.3.5 Cropping

Strip cropping. Strip cropping is where two different crops are planted in alternate strips, or where three or more crops are planted sequentially in strips, usually on the contour or at least across the slope (or perpendicular to the direction of prevailing winds in areas where wind erosion is a concern). The width of strips ranges from narrow, *i.e.,* a few meters, up to 20 meters or more. The potential of strip cropping to reduce pesticide runoff losses results from 1) possibly decreasing pesticide use (if one of the crops does not require the pesticide of concern) and 2) decreasing pesticide transport where the untreated crop could act as a buffer strip (discussed later) and/or reduce erosion, runoff, and sediment transport (particularly if the crop is a close-grown crop acting like a vegetated filter strip [also discussed later]). The potential advantage of reducing pesticide runoff losses is likely greater for more strongly adsorbed pesticides. With a 50% reduction in application and with equal alternating areas of treated and untreated crops, losses could be reduced up to about 75% over a monoculture treated crop. The feasibility of strip cropping is dependent on soils, slopes, irrigation capability of crops, etc.

Other advantages include reductions in other environmental pesticide losses as well as reduced inputs and reduced erosion. Crop yields in some cases may be increased as well. Possible disadvantages include more management time spent keeping track of the locations of the strips and care in use of some pesticides on one crop if the adjacent crop is susceptible to pesticide damage. In addition, if narrow strip cropping is used, the beneficial rotation effect may be reduced if pests move from one strip to another. For example, corn rootworm adults could move from soybean strips (rotated from corn the previous year in a corn-soybean strip-cropped field), requiring an insecticide treatment.

Crop rotation. The use of crop rotation (*e.g.*, corn following soybeans) has considerable potential for reducing pesticide runoff losses by decreasing or eliminating the need for some pesticides. For the example of corn following soybeans, the need for a rootworm insecticide is much reduced over corn grown every year. There may also be a beneficial effect of reduction in soil and water losses as well where a rotation into a sod crop will reduce erosion and runoff while that crop is being grown, as well as having a carry-over effect to at least the first year out of sod. With total elimination of pesticide use, loss would be reduced 100%.

In addition to reduced pesticide losses to other resources, crop rotation often has the advantage of increased yields, with all other inputs being equal. One disadvantage is that economics due to need (or lack of need) for certain crops or farm programs do not allow the flexibility required to rotate crops (also true for strip cropping).

In addition to economics, including government incentives, the feasibility of crop rotations is determined by climate and soil factors.

4.3.4 Management Factors—Field-to-Stream Transport Reduction

Structural practices, such as terraces, can reduce field-to-stream transport of pesticides by reducing sediment transport and runoff. Likewise, landscape reconfiguration, including the use of buffer strips and wetlands within watersheds or as riparian zones, can potentially attenuate the amount of pesticides carried with sediment and dissolved in water upon passage through these features.

4.3.4.1 Structures

Terraces/detention ponds. Terraces and detention ponds are used as a method of erosion control by constructing broad channels or dams across the slope of rolling land. The major type of terrace of economic importance used in the U.S. is the broad-base terrace. In constructing the channel, an embankment is usually constructed to either allow water to slowly drain down a more nearly level channel (graded terraces) or to dam the water and force infiltration (level terraces) or provide an outlet from the detention pond (tile-outlet terraces). Because of construction costs and the desire to avoid crop damage, most terraces built today are of the tile-outlet (or detention pond) type. As such, they are designed to release the runoff (*e.g.*, about 5 cm in Iowa) from a 10-year 24-hour return interval storm over a 48-hour period through an underground tile system. In that period, infiltration might reduce runoff some, but the major effect is the reduction of sediment load in the outflow (usually by 80 to 95%) by deposition in the pond or terrace basin. Thus, the predicted effect of the terraces on pesticides transported with sediment and runoff would likely range from 5 to 90% reduction, depending on how much of the pesticide is adsorbed or transported with sediment.

Other advantages of terraces are obviously related to the effects of reduced erosion on the land and reduced sediment (and nutrient) transport to water resources. In addition, depending upon the water retention time of the ponds and the route of degradation, detention ponds may also result in greater degradation of the pesticide. This may be especially true if the pesticide of concern undergoes rapid hydrolysis.

The concerns with terraces are their initial and maintenance costs, their effect on the "farmability" of the land, and the land they may take out of production (with steep slopes, push-up terraces are often constructed with the steep outer face left in grass).

The effectiveness of terraces on the reduction of field-to-stream transport of pesticides is closely tied to the soil adsorption properties of the pesticide of concern. For strongly adsorbed pesticides, the effectiveness will be much greater than for weakly or moderately adsorbed pesticides. For the particular case of tile-outlet terraces and a weakly or moderately adsorbed pesticide like atrazine, there is concern that direct transport to a stream from the field through underground tile eliminates any possibility of attenuation in the transport process. The atrazine label was rewritten in 1993 to require a 66-foot buffer around either the inlet or the outlet to help alleviate this concern.

Constructed wetlands. Constructed wetlands are designed to include saturated substrates, emergent and submergent vegetation, animal life, and water that simulate natural wetlands. Constructed wetlands have the potential for removal of pesticides dissolved in water and associated with sediment in surface runoff. In general, wetlands are expected to be effective in removal of suspended particles and pesticide attached to them and not nearly so effective in removing pesticides dissolved in water. However, little quantitative information is currently available to better assess these potentials. Of particular importance will likely be the detention or travel time for flow through these impoundments or wetlands. The effect of wetlands on pesticide runoff losses might range between 0 and 90%.

Other advantages of constructed wetlands might be nutrient (particularly nitrate) removal from agricultural drainage and possibly the use of the wetlands as a source of surface water for some needs. Concerns include the cost of construction and maintenance and land taken out of production. In addition, the wetlands themselves may be considered important ecological areas that require protection. If growers construct wetlands for runoff mitigation, then the concentrations of pesticides in the wetlands should not be regulated the same as "natural" surface waters.

Pesticide properties, particularly soil adsorption and degradation modes, will likely be major factors in determining how effective wetlands can be in the reduction of field-to-stream transport of pesticides.

4.3.4.2 Landscape reconfiguration

Buffer strips. A buffer strip is an area on a field border that is not treated with pesticide. It can be an area that is being planted to the same crop as that within the field, in fallow, or be planted to some other crop (if planted to grass or some other close-grown forage crop, it will be considered a vegetated filter strip). The purpose of the buffer strip is generally three-fold: first, it can physically separate a resource of concern (*e.g.*, a stream) from a treated area; second, it reduces the amount of pesticide used in the watershed (however, usually by only a few percentage points); and third, if overland flow from the treated area passes over the untreated area, there is the potential for attenuation of the pesticide concentration carried in the surface runoff by adsorption or deposition. Infiltration in the buffer strip may also reduce runoff volume. Overall, the expected reduction in pesticide transport would likely range between 5 and 40%.

Other advantages would include reduced risk to the groundwater and the atmosphere with reduced pesticide application. Concerns for efficiency of the practice would involve relative areas of the contributing watershed and the buffer strip (the larger the ratio, the lower the efficiency) and the path of water flowing through the buffer strip (the more concentrated the flow, the less interaction with materials in the buffer strip and the less the attenuation). Economic disadvantages could come with decreased pest control (for the case where the buffer strip is in the same crop), decreased crop area (for the case where the buffer strip is kept fallow), and decreased crop production (for the case where the buffer strip is in another less productive crop or one susceptible to the pesticide in runoff from the treated area).

Other major factors influencing the predicted effect include how strongly a pesticide is adsorbed. Transport of less soluble, more strongly adsorbed pesticides will be reduced the most with sediment deposition and/or runoff interaction with in-place soil. In a similar vein, a combination of climate, hydrologic, and soil factors resulting in more erosion could make this practice more effective.

Mixing/loading/handling set-backs. These set-backs are distances that must be maintained from a point of mixing/loading/handling to a water resource of concern such as a stream, lake, reservoir, or groundwater well. With the exception of some inconvenience in certain instances, there is little cost to this method of potential prevention of catastrophic results from spills.

Vegetative filter strip. A vegetative filter strip is a buffer strip planted to grass or some other close-grown plants, normally of a forage type (but might include shrubs or even trees). As with the buffer strip, the purpose of the vegetated filter strip is to remove pesticides in solution or associated with sediment from runoff by filtration, deposition, infiltration, adsorption, decomposition, and/or volatilization. By both slowing runoff velocity and providing more biological surface area (living and dead) for interaction, the vegetative filter strip is expected to be somewhat more efficient in reducing the field-to-stream transport of pesticides, likely in the range of 10 to 60%.

Other advantages and concerns are similar to those for buffer strips. Specifically, to be effective, runoff must not concentrate or channelize, but ideally must pass through the vegetation in nearly uniform sheet flow. The vegetation must be erosion- and pesticide-resistant. The lower the ratio of contributing watershed to filter strip area, the longer the contact time and the greater removal efficiency.

Removal efficiency depends on pesticide properties, with less soluble, more strongly adsorbed pesticides likely to be more affected. In addition, as with buffer strips, climate, hydrologic, and soil factors resulting in more erosion could make this practice more effective.

Grassed waterways. Grassed waterways are in effect vegetative filter strips with a somewhat different orientation relative to inflow and outflow directions and a somewhat different purpose. Grassed waterways are generally designed to collect overland flow and then transport it down gradient to a drainage outlet without erosion within the waterway. As such, the grassed waterway acts as a vegetated filter strip during the initial entry of runoff into it, with decreasing function as a filter as flow concentrates in the lower elevation of the channel or waterway. Because of generally much greater ratios of watershed to grassed areas, the likely range of field-to-stream transport reduction is 2 to 40%, less than either buffer strips or vegetative filter strips.

Other advantages are the same as those for buffer or filter strips including erosion control. The concerns are also similar but with a particular concern for sediment deposition in the outer edges of the grassed waterway causing a rise in elevation resulting in a "dam." This dam prevents water from entering the waterway and causing concentrated flow and erosion in the unprotected soil bordering the waterway. Therefore, maintenance of the waterway and erosion control in the field above the waterway are both necessary.

4.3.5 Summary of Runoff Mitigation Options

Table 4-2 provides a summary of all of the mitigation options discussed for runoff and provides the range of reductions expected and specific comments associated with each option. An important point to be remembered is that within these categories, the percent reductions are not necessarily additive. Implementing one option can possibly increase or decrease the percent reduction that may be obtained for implementing other options.

TABLE 4-2. Mitigation Practices Summary Guide* for Pesticide Runoff Losses to Surface Waters

Practice	Potential Reduction of Surface Runoff Transport**		Comments
	strongly adsorbed***	weakly to moderately adsorbed	
Field Loss Reduction:			
lower application rate	0-50%	0-50%	loss reduction should be ≥ rate reduction; *e.g.* , at 3/4 rate, loss should be reduced at least 25%
partial substitution	0-80%	0-80%	environmental concerns may also exist for pesticide(s) used as substitute(s); upper range would go to 100% with total elimination of use
partial treatment	0-75%	0-75%	*e.g.* , herbicide banding; loss or reduction in pest control and/or alternative treatments must be considered
formulation	0-25%	0-50%	potential effects need to be documented in field, laboratory, and/or modeling studies
soil erodibility/special restrictions	0-50%	0-25%	restrictions should be targeted to more strongly adsorbed pesticides used on highly erodible land
soil incorporation	25-50%	35-70%	mechanical incorporation reduces the amount in surface mixing zone; more important for solution losses
application timing	0-50%	0-50%	loss decreases with time between application and storm-runoff; probabilistic weather information could be used
no-till	50-90%	0-40%	erosion control by 90% feasible; runoff volume reduction much less; herbicide wash off from residue may increase concentrations in runoff
conservation-tillage	40-75%	0-50%	erosion control less than for no-till; runoff reduction for first storm after application more reliable than for no-till
subsurface drainage	0-20%	0-50%	subsurface drainage can reduce antecedent moisture and therefore runoff and erosion; infiltration can reduce surface concentrations for less strongly adsorbed pesticides
avoid sealing/compaction	0-20%	0-50%	very similar to the effects of infiltration differences caused by subsurface drainage
irrigation	0-25%	0-50%	improved management practices reduce runoff and erosion; greater infiltration could reduce concentrations for less strongly adsorbed pesticides
strip cropping	0-75%	0-60%	possible combination of reduced use (untreated strips) plus buffer effect (sediment deposition on contour)
crop rotation	0-90%	0-90%	pesticide needed could be much reduced in some rotations
Field-to-Stream Transport Reduction:			
terrace/detention ponds	20-90%	5-20%	sediment transport reduction; infiltration in basins could reduce runoff volumes and therefore losses
constructed wetlands	20-90%	0-50%	a practice for which little quantitative information exists
buffer strips	10-40%	10-25%	relative area untreated to total area important; assumed to be ≤ 10%
set-backs	0-50%	0-25%	protection from spills (point-source) during mixing/loading/handling
vegetative filter strip	20-60%	10-40%	to be effective, runoff must pass through at nearly uniform depth; removal more efficient for lower contributing area-filter strip area ratio
grassed waterways	10-40%	2-10%	similar to filter strip, but likely with higher contributing area-filter strip ratio; concentrated flow reduces effectiveness

* The rough estimates of the likely range of effects for each practice are based on limited research and/or professional judgment.

** It should be possible to predict a more narrow range for potential reduction using mathematical modeling for a specific pesticide and a specific set of soil and environmental conditions.

*** Partition coefficient, or K, typically > 100

4.4 APPROACHES FOR MITIGATING DRIFT

The purpose of this section is to provide a brief overview of factors to be considered when trying to minimize or reduce off-site movement of chemicals via drift. The rationale for the use of specific mitigation measures, along with the primary advantages and disadvantages of the various approaches, will be provided. In some cases, the expected range of percent reductions that may occur will also be provided; however, for the most part, quantitative data are lacking.

The major spray application scenarios where drift can occur are

- ground,
- chemigation,
- airblast, and
- aerial (fixed wing and rotary).

Of these, two of the application patterns are of primary concern:

1. aerial applications and
2. airblast applications.

Many of the factors to be considered in reducing drift are based on similar principles and are often directly applicable to either aerial or airblast applications. Other factors can be considered specific to each use pattern. A summary of drift mitigation factors can be found in Table 4-3.

Many of the aspects associated with drift are currently being evaluated by the Spray Drift Task Force (SDTF). The SDTF is an independent, joint development project of pesticide registrants, formed in April of 1990 to develop pesticide spray drift data in support of task force members' product registrations and reregistrations. Working closely with USEPA, the SDTF is developing a generic spray drift data base that quantitates drift from aerial, airblast, ground boom, and chemigation application scenarios under a variety of environmental conditions. The drift data base is scheduled to be completed in 1994/1995.

TABLE 4-3. Mitigation Practices Summary Guide for Pesticide Spray Drift Losses

Mitigation Factor	Rationale and Comments
Buffer Zones or Set-Backs	acts as intercept area for swath displacement and drift; percent reduction in drift will depend on size of buffer area
Meteorological Related Restrictions	
Wind Speed	in general, drift increases with wind speed; restrictions on applications at high wind speeds could decrease drift by 5-40%
Wind Direction	if wind not blowing towards sensitive area of concern, then drift into area of concern will not occur; restrict application only when wind blowing toward sensitive area
Temperature Inversions	drift can increase during inversions; restricting application under such conditions should decrease drift; percent reductions attainable unknown
Humidity	droplet evaporation will be greater when humidity is low; restricting applications during low humidity will decrease drift; percent reductions unknown, likely in 10-20% range
Equipment Specification	
Nozzle Type	requiring nozzles that produce coarse droplets will reduce drift; percent reductions unknown and will vary based on specific nozzles
Nozzle Orientation	nozzles oriented into the airstream will produce a greater percentage of fine droplets; minimization of nozzle angle will reduce drift; percent reduction unknown
Boom Configuration and Length	minimizing boom length can decrease wing vortex effects and decrease drift; 10-20% reduction in drift may be attainable
Pressure and Flow Rate	lower pressure can minimize formation of fine droplets and thus reduce drift; related to other equipment factors; percent drift reduction fairly minor
Volume	equipment configurations required to apply at low volumes result in the formation of finer droplets; restricting low volume use can decrease off-site movement by 10-100%, depending on the base case
Specialized Equipment	for tree crop applications, specialized equipment that can reduce drift is available ("sensor" sprayers ,etc.).; percent reductions in drift attributable to such equipment are unknown
Application Procedures	
Applicator Speed	aircraft speed influences shear across nozzle; higher speed - greater shear - more fine droplets; restricting applications at higher speed may reduce drift; percent reductions unknown
Height Above Canopy	greater height above canopy - greater drift potential; restricting maximum height can reduce drift; percent reduction unknown
Matching Sprayer to Canopy	for airblast sprayer, tree size critical and drift can increase if equipment not matched to tree geometry; percent reductions unknown
Application Timing	drift can be minimized by applying during "low drift" periods; "low drift" periods generally associated with favorable meteorological conditions
One-Way Spraying	for orchard spraying, spraying only toward the center of the orchard and away from the sensitive areas may decrease off-site movement; percent reduction unknown
Tank Additives	various tank additives are available, which claim to minimize the formation of small droplets and/or decrease evaporation rate; limited data available confirming actual reductions in drift
Site Modifications	
Wind Breaks	structures (i.e ., trees) can be planted either adjacent to the habitats of concern or around the treatment area (i.e, orchard) to intercept drift; minimal quantitative data available on interception effectiveness; long-term option.
Conversion to Smaller Trees	for tree crops, drift is more controllable when available to small trees; long-term option.

4.4.1 General Factors for Drift Reduction

4.4.1.1 Buffer zones

A commonly used approach for minimizing the exposure of nontarget areas to chemical drift is the imposition of a buffer zone or set-back area. As discussed for runoff, a buffer zone is an area of defined width situated between the target application area and the nontarget "area of concern." Relative to drift reduction, this area can intercept swath displacement not entirely accounted for via off-setting during application and, based on the width, can intercept a large percentage of off-target movement. The width of the area required depends on the reduction necessary to achieve an exposure below the LOC.

Theoretically, buffer zones will always be capable of reducing exposure below the LOC. The key to setting the most appropriate buffer zone width is balancing the need to reduce exposure to a reasonable and acceptable level, while considering the practical implications of buffer zones for the grower. Relatively small buffer zones can reduce exposure significantly. For example, for low volume (< 2.5 gal/A) aerial applications, a buffer zone of 75 feet can give an additional 50 to 75% reduction in the amount of pesticide deposition in nontarget areas.

Factors to be considered when evaluating buffer zones include

- percent reduction necessary to achieve the desired goal,
- typical application parameters associated with the crop use,
- regional meteorological considerations,
- practical aspects of buffer zone restrictions given the crop and region, and
- vegetative requirements for the buffer zone area.

The primary advantage of buffer zones is that they result in a direct, quantifiable reduction in off-site deposition. Buffer zones represent an enforceable exposure-reduction mechanism which is easily understood and which growers can implement without modifications in equipment or use practices. The main disadvantage is the potential economic impact associated with taking land out of production or yield reductions from untreated acres (lack effective pest control). The relative importance of these factors will vary with crop and region. For large acreage crops, such restrictions may not impose undue burden on the growers. However, for smaller acreage crops, a buffer zone can significantly impact the available acreage for planting. Such economic factors need to be considered on a case-by-case basis.

The size of the buffer zone required will depend on the other operational parameters that determine the droplet size spectrum for a particular use scenario (discussed in greater detail below). If equipment specifications are such that large droplets are produced, then buffer zone requirements should be less than for scenarios which produce finer droplets.

When implementing buffer zones, it is important to clearly articulate the details on the label. For example, an important consideration is the cropping practice required or permitted within the buffer zone. The specific details of what is permitted in the buffer zone may vary on a case-by-case basis. For example, can the crop be grown in the buffer area?

If it is grown, the crop within the buffer area would not be treated with the specific chemical of concern; however, other methods of control could be used. With this approach, cropland is not required to be taken out of production. Also, with some type of cover in the buffer area, additional drift reduction can be attained via crop interception. However, two main problems with retaining the buffer area in production are

1. the buffer area can act as a source for pest species re-infestation and
2. with the crop present, enforcement of the label restriction is more difficult.

The relative reduction in drift from the implementation of buffer zones should be quantifiable. The SDTF is currently conducting an extensive series of field trials which will enable fairly accurate estimates of the reductions in drift associated with specific buffer zone widths. The deposition curves from these trials will enable initial estimates of potential reductions for various size buffer zones. If the initial estimates do not provide the adequate accuracy, spray drift models can be used to refine the estimates.

4.4.1.2 Meteorological conditions

Weather conditions during and immediately after application directly influence drift potential. The importance of weather conditions will vary based upon other application parameters; however, a number of specific factors exist which, if accounted for, can help to minimize off-site movement. The primary factors are wind, temperature, humidity, and atmospheric stability.

Wind direction. If the wind direction is not toward the habitat of concern, exposure should be minimal. Therefore, imposition of specific mitigation measures (buffer zones, wind speed restrictions) may only be necessary when the wind direction is toward the habitat of concern. For example, a statement on the label such as, "Do not apply this product when wind speeds are greater than 10 mph and the wind direction is 'perpendicular' to any aquatic habitat of concern (within 45%)," may be considered acceptable mitigation. Such a restriction may be most effective in combination with a buffer zone.

The advantage of such a mitigation approach is that it allows the grower the greatest flexibility in that mitigation is only implemented when conditions favor movement toward the aquatic habitat of concern. The disadvantage of such an approach is effective implementation and enforceability, specifically,

- application sites may be on more than one side of the habitat. This may make it very difficult to determine when and where a grower can apply a product.
- wind direction can shift considerably in relatively short time periods. With such shifts, determining the area that can be treated would be difficult.

Although these factors may limit the usefulness of this approach, this still may represent a potential option for certain regions of the country or for specific use patterns.

Wind speed. Wind speed can have a direct influence on off-site movement, although the relationship is not necessarily linear. Since wind speed is basically a function of other weather variables that influence drift, such as temperature and humidity, it is not always possible to directly relate wind speed restrictions to reductions in drift. Theoretically, the greater the wind speed, the lower the within-swath deposition; however, such a reduction in on-target recovery does not always relate to greater off-site deposition, as in these examples:

- At very low wind speeds (<1 to 2 mph) temperature inversions often exist. The temperature boundary created by an inversion can result in fairly high off-site movement, although near-site deposition may not necessarily increase because of the lack of turbulence, which is necessary for small droplets to deposit.
- Depending upon the crop canopy and turbulence, higher wind speeds may actually increase deposition. Some degree of turbulence is required for deposition and may actually increase deposition of smaller droplets in the target area.

There is little doubt that restricting applications at very high wind speeds (>20 mph) will reduce the potential for off-site movement. However, quantitative data on the exact reductions associated with such restrictions are not available. Literature data are too variable to quantitatively evaluate the reductions in drift attributable to wind speed. Many studies imply that wind speed is critical; however, none truly quantitate the impact of wind speed on drift. In most cases, it is likely that wind speed primarily influences swath displacement of the larger droplets and does not drastically alter the drift of the smaller droplets. The initial results from the SDTF indicate that labels restricting aerial applications to wind speeds of less than 15 mph could reduce drift by 10 to 20%. Often, states and aerial application handbooks recommend that applications should not be made in winds less than 3 mph or greater than 10 mph. However, such a restriction is not always possible or necessary and should be evaluated on a case-by-case basis.

Advantages of wind speed restrictions are that they are easily understood, logical, and fit within the training and current understanding of drift control practices. Exact measurement of wind speed can be difficult, especially since wind speed will vary greatly with height, although an understanding of the wind conditions at the time of application is essential for safe operation of an aircraft, and thus the information can be considered generally available. Overestimates of wind speed are often obtained if one relies on airfields for estimates, since most airfields measure wind speed at heights (30 feet) significantly greater than typical application heights. For the most part, wind speed is an enforceable label statement, assuming that inspectors/regulators have the appropriate equipment or experience. Additional data are required to be able to quantitate the importance of this variable, and such data are currently being generated by the SDTF.

Temperature inversions. As defined earlier, meteorological conditions during temperature inversions can increase the off-site transport of small droplets. However, the increased off-site movement is generally related to longer range transport and will not necessarily increase near-site deposition. Because of the greater loss potential, it is generally recommended that applications not be made during temperature inversions. Inversions can occur in any geographic region, although there are specific local conditions (coastal areas and specific terrains such as valleys) which can be identified as particularly prone to the formation of an inversion. These situations can be controlled through specific label restrictions.

Although it is known that off-site movement can be greater during inversions, there are limited data that can quantitatively predict how much greater deposition in near-site habitats would be during inversions. Comparative studies that maintain application variables constant and simply evaluate drift under stable (inversion) and unstable atmospheric conditions have not been performed. So from a mitigation perspective, it is not possible to quantitatively predict what decrease in exposure may result from a label statement that restricts applications during temperature inversions. At this point, it is unknown how well models currently being evaluated by the SDTF will address drift during inversion conditions.

As an enforceable mitigation measure, the greatest difficulty associated with temperature inversions is knowing when an inversion exists. Aerial applicators can use a temperature probe located outside their aircraft to measure the temperature gradient and determine whether an inversion is present. Smoke (from a smoke pot or a burning tire) is another mechanism to evaluate local weather conditions. If an inversion exists, vertical movement of the smoke will be restricted by the warm overhead ceiling or boundary. Such approaches for measuring for inversions exist; however, they are not applied consistently by applicators. Greater education and training are required in this area of drift control.

The advantage of restricting applications during inversions is that conditions which favor off-site movement can be avoided. However, at this point, it is unknown whether this will truly reduce the exposure to nearby aquatic habitats. If the only reason for such restrictions is to reduce aquatic exposure, it may be premature to impose such measures because of the difficulty associated with determining when an inversion exists and thus the increased hardship on the applicators or growers. Greater standardization of procedures for pesticide application under such conditions needs to be developed.

Temperature and humidity. Temperature and humidity are closely interrelated with other meteorological variables that directly influence drift. Atmospheric turbulence and associated wind speeds are primarily controlled by temperature differences and, as discussed earlier, can directly influence drift. Apart from these indirect effects, temperature and humidity can directly influence drift by altering the evaporation rate of small droplets. After atomization, the droplet spectrum is subjected to varying environmental conditions. The larger droplets (greater than 200 μm) generally settle before their size can be drastically altered by evaporation. However, for smaller droplets, evaporation can significantly decrease the size of the droplet and thus increase drift potential. Temperature and humidity can directly influence the rate of evaporation. Under hot and dry conditions, evaporation will be extremely rapid. Under cool, moist conditions, evaporation will be slowest. Evaporation is only a factor when the spray solution is water based. When oils are sprayed, evaporation is not a consideration. In addition, the formulation and percent volatiles in the tank mixture, and the use of tank additives, may decrease the droplet evaporation rate.

Temperature and humidity are certainly not the most important factors influencing drift potential; however, they could be very useful mitigation approaches in certain geographic regions. In hot and dry climates, the humidity can be as low as 8 to 10% at midday. Restricting application during such time periods may reduce deposition in nearby aquatic habitats.

Data for predicting the potential decrease in off-site deposition that may be associated with such a restriction are not readily available. Simulation models have shown that decreases may be expected; however, these models are not completely validated. The SDTF is specifically evaluating the influence of humidity and temperature on drift, and quantitative data should be available in the near future.

Restrictions based on temperature and/or humidity are easily understood, measurable, and enforceable. The disadvantage of such a restriction is that in hot and dry areas, such a restriction may only allow a small window of time for application. Given that there may not be extensive aquatic habitats in such areas, buffer zones used in combination with practices that minimize the formation of finer droplets may represent a more reasonable drift mitigation approach.

4.4.1.3 Regional use restrictions

Regional use restrictions are primarily predicated on variations in local equipment specifications, application procedures, cropping practices, or meteorological conditions and may offer unique opportunities for drift mitigation. Conversely, the same conditions may represent reasons why specific mitigation factors may not be practical for certain regions. Regional considerations should not only focus on mitigation measures, but should also address whether the aquatic habitats of concern vary regionally.

Specific restrictions that may be considered include, but are certainly not limited to, the following:

- Hot and dry climate—Specific regions, such as the desert Southwest, may have weather conditions conducive to the formation of small droplets.
- High winds—Some geographic areas can have consistently high winds during much of the growing season and thus may require special considerations.
- Temperature inversions—Specific regions or conditions may be particularly prone to the formation of inversions.
- Ultra low volume (ULV) applications—Such applications predominate in specific regions, and this should be considered in evaluating mitigation options.

The advantage of regional use restrictions is that they address a specific problem and the conditions associated with the problem. Any mitigation effort should focus on the primary areas where the concern exists and should evaluate the specific factors controlling the exposure and associated with the use of the product under those conditions. The advantages of this approach are that it will not place undue hardship on those regions where the problem does not exist, and regional use restrictions can also focus enforcement efforts in specific regions.

4.4.2 Drift Mitigation Approaches for Specific Application Scenarios

The preceding discussion can be directly applicable to most application scenarios. However, there are other factors that are specific to the use patterns being evaluated (ground, aerial, chemigation, airblast). Specific application scenarios are addressed below. Factors discussed above will be reviewed if additional information specific to the application scenario is relevant.

4.4.2.1 Ground application

In the majority of cases, application by ground hydraulic sprayers does not result in significant off-site drift. Under some circumstances (low volume, high boom conditions) drift can occur; however, it is generally an order of magnitude less than drift from aerial applications. Small buffer zones (10, 25, 50 feet) and/or optimization of equipment to minimize formation of small droplets are generally adequate to account for any swath displacement or drift. Factors discussed above, concerning meteorological conditions, and below, concerning equipment specifications, are generally applicable to ground application scenarios.

4.4.2.2 Chemigation

Based on the extremely high volumes, low pressures, and fairly large droplet spectrums associated with application of chemicals through irrigation systems (chemigation), one would theoretically predict fairly low off-site movement. At this point, however, there are limited data available on drift from chemigation. The primary equipment specification that could result in off-site movement is the use of end guns. The SDTF is currently evaluating results from chemigation field studies and, based on these results, the necessity of drift mitigation measures should be evaluated.

4.4.2.3 Aerial applications

4.4.2.3.1 Meteorological conditions

All of the discussion in Section 4.4.1.2 on Meteorological Conditions is directly applicable to aerial applications. Since wind speed will vary with height above the canopy, this factor is especially critical for aerial applications. For example, if wind speed is five mph 2 to 3 feet above the canopy, speed may be ten mph 20 feet above the canopy. In addition, since spray atomization set-ups for aerial application can produce extremely fine droplets, meteorological factors such as wind, inversions, and humidity can be critical in determining the off-site movement.

4.4.2.3.2 Equipment specifications

The specific equipment used to apply pesticides by air can directly influence drift and thus must be considered when attempting to control off-site movement. Many of the options available directly influence the droplet spectrum produced, and although discussed individually below, should be considered in combination.

Nozzle type. Atomization of the spray solution occurs as the solution passes through the application nozzle. The specific characteristics of the nozzle and the wind shear across the nozzle (in addition to the pressure, flow rate, and physical properties of the solution) will determine the droplet spectrum produced during atomization. The mechanism of atomization differs among nozzle types (hydraulic, rotary [micronair], air assist, etc.) and influences both the droplet size and distribution. Numerous research projects have investigated and documented the droplet spectrums for different nozzles under different conditions. These data, along with the ongoing atomization work of the SDTF, form the foundation for recommending specific nozzle types for reducing off-site movement.

Although there are considerable data on the droplet spectrum associated with various nozzles, specific field data for aerial application which quantitatively relate a reduction in the small droplets via changes in nozzle type with reduced off-site deposition are lacking. Simulation models can be used to estimate the drift reductions possible by switching to "coarser" nozzles; however, quantitative field confirmation is presently limited. It is fairly safe to assume that requiring the use of coarse nozzles (or restricting the use of nozzles that produce very fine droplets) will reduce off-site movement; the unknown is quantitative estimates of the level of reduction possible. The SDTF is currently developing the field and atomization data to make such predictions possible.

Since a vast array of nozzles is currently available, requiring specific nozzle types could present difficulties for aerial applicators. In addition, applicators are often applying numerous products within relatively short time frames. Having different nozzle requirements for different products may not be practical. If nozzle-specific mitigation measures are to be implemented, they will need to be flexible and fairly straightforward. As opposed to stipulating specific nozzles, it may be best to stipulate general practices that will result in a droplet distribution within a specified range.

Nozzle-specific restrictions should be enforceable based upon inspection of either the set-up of the aircraft prior to application or records of the application parameters for specific products.

Nozzle orientation. The wind shear across the nozzle will also influence the droplet distribution. Wind shear is determined by both the speed of the aircraft and the orientation of the nozzle in the airstream. Nozzles oriented at 90 degrees to the airstream will produce a larger percentage of fine droplets than those oriented at 0 degrees (straight back). Since better coverage and canopy penetration can be obtained with smaller droplets, especially for low and ultra-low volume applications, applicators may consider it desirable to orient nozzles further into the airstream. Unfortunately, this arrangement also produces greater drift potential. For drift mitigation purposes, recommending a nozzle orientation of 0 degrees or restricting applications at 90 degrees may be viable options.

As with nozzle type, this mitigation approach is based on the concept that practices which minimize the formation of small droplets should reduce off-site movement. Comparative field data in which all other conditions are equal and the influence of nozzle orientation was evaluated are not currently available. It is likely that a combination of nozzle type and orientation to produce the largest droplet spectrum, while maintaining practicality and efficacy, is most appropriate. The SDTF is currently developing data that should allow the evaluation of nozzle orientation as a mitigation factor.

As with nozzle type, nozzle orientation is an enforceable option which should be considered. The advantages of nozzle orientation are that compliance with the requirement is fairly simple and significant reductions in drift may be obtainable.

Nozzle configuration and boom length. Nozzle position on the boom will not influence the overall droplet spectrum; however, for the smaller droplets produced by the outer nozzles (closest to the wingtips), the influence of the wingtip vortices can be important. Entrapment of the small droplets in the vortex can cause the upward movement of droplets into the airstream, resulting in greater potential for off-site movement. Elimination of the outer nozzles closest to the wingtip will not greatly decrease the effective swath width, but it will reduce the entrapment of the smaller droplets in the vortex. Booms should be configured and patterns tested to obtain a boom configuration that minimizes such vortex effects.

Limited data are available on the relationship between boom length and off-site movement of chemicals. Preliminary results from the SDTF indicate that decreases in drift of 10 to 20% may be observed by reducing the boom length from 87% to 66% of wingspan. A common recommendation in aerial applicator handbooks is that the boom length should not exceed 75% of the wingspan. Although such a recommendation is certainly reasonable and "directionally correct," a difficulty with such generalizations is that a tremendous number of different aircraft and/or boom configurations exist, and the importance of wingtip vortex effects, and thus the reductions in drift obtainable by restricting the boom length, will vary from plane to plane.

In general, boom length restrictions are considered an enforceable mitigation option. Based on current knowledge, quantitation of the decreases associated with such restrictions is difficult, and the relative reduction obtained will vary with the type of plane and boom.

Application pressure and flow rate. The pressure and flow rate for the nozzle can also affect the droplet size distribution. For hydraulic nozzles, which rely on pressure for the atomization of the solution, greater pressures can result in the formation of smaller droplets, which in turn may result in greater drift, although this may not always be the case. To reduce drift potential, it is desirable to minimize system pressure. Greater flow should be obtained by switching nozzle size or increasing the number of nozzles, not by increasing pressure.

Pressure limitations for specific uses may provide a viable mitigation measure; however, the exact reductions in expected drift are currently unknown. A problem with this approach is the enforceability of specific pressure restrictions.

Application volume. Directly related to application pressure and flow rate is the application volume. Low volume (LV; 0.5-2.5 gal/A) and ultra-low volume (ULV; <0.5 gal/A) applications are extremely cost effective for the applicators (decreases the number of spray tank refills required); however, LV and ULV generally produce a larger number of small droplets. ULV applications are normally made using crop oil as the diluent/carrier in place of water (although diesel fuel has also been used), whereas LV applications normally use water as the diluent. With the use of oil as the diluent, concerns regarding post-atomization evaporation are eliminated. Given the extremely low volumes associated ULV, the generation of smaller droplets is often necessary to obtain adequate coverage and canopy penetration. Unfortunately, the drift associated with the smaller droplets can be a problem. As a risk mitigation issue, the objective is to recommend the volume (and other associated equipment variables) that provides the largest droplet distribution without affecting the efficacy of the product.

Data available in the literature and generated thus far by the SDTF indicate that drift associated with ULV applications can be considerably greater than with higher volume applications. The exact reductions which could be attributed to prohibiting ULV applications are unknown at this point, although the SDTF will be generating enough data for such predictions. It is expected that such restrictions could significantly reduce off-site movement. Prohibiting ULV applications under specific circumstances represents an enforceable mitigation measure that could alleviate some concerns regarding drift; however, the economic impact would need to be carefully considered on a case-by-case basis.

4.4.2.3.3 Application procedures

Specific application procedures to be considered as mitigation measures include the speed of application and the height above the canopy during application.

Air speed during application. The wind shear across the nozzle orifice can directly influence the break-up of the atomized spray solution and thus the droplet size distribution. Typical application speeds in the United States for fixed-wing aircraft range from 75 to 150 mph. For helicopters, application speeds are considerably lower, ranging from 10 to 60 mph. As application speed increases, the formation of smaller droplets also increases. Restrictions on application speed may minimize the formation of smaller droplets, thus decreasing potential drift. Atomization studies conducted in wind tunnels have shown an increase in the number of fines as application speed increases; however, at this point, comparative field data do not exist to show the percent increase in drift associated with such a reduction in drop size. In the case of aircraft speed, it is not only the change in droplet size that influences the differences in drift. Other equipment and application factors associated with higher application speeds, such as aircraft wake effects and height of application, may also influence the actual drift observed under field conditions.

Restrictions in application speed offer a potential drift mitigation factor that may be directly quantifiable. At this point, the percent reductions attainable for off-site movement are unknown. In addition, an evaluation of the economic impact of such restrictions needs to be considered. Higher speed applications (*i.e.*, turbo planes) enable applicators to cover considerably greater area in a given time period, and this results in lower costs to the growers.

Application height above canopy. The height of application above the canopy can significantly affect off-site movement. There are four points to consider relative to application height and drift.

1. The greater the application height, the greater the settling distance before impacting the crop or target. Given a specific crosswind, this allows greater off-site displacement of the droplets.
2. For the smaller droplets, evaporation can be rapid. Longer residence time in air will result in the formation of smaller droplets via evaporation and thus greater potential for drift.
3. As height above canopy increases, wind velocity increases. Greater drift may result at the greater wind velocities.
4. Pilot safety may be reduced at lower application heights.

Given these factors, height above canopy should be minimized. For typical agricultural crop applications, height should be between 6 and 12 feet above the canopy, based on the terrain and other canopy factors. Application height restrictions may reduce off-site movement, although enforcement could be a problem. The SDTF is currently evaluating application height as a variable, and preliminary results indicate that height can be an important factor. The percent reduction attributable to such a restriction is unknown and may be difficult to quantify, since the base case application height is not precisely known and will vary substantially.

4.4.2.3.4 Canopy effects

The presence of a canopy can potentially influence off-site movement of products through modifications in the micro-climate and direct interception. Drift from bare ground fields is generally considered to be greater than drift from applications made to full canopies. However, the influence of canopy on drift does not offer any specific mitigation factor that can directly reduce off-site movement. The timing of application (to ground, partial, or full canopy) is generally a function of the target pest and not necessarily controllable. In cases where products are applied both early season and later in the year (full canopy), the potential difference in drift is unknown. More research is needed to determine the exact importance of canopy in off-site drift. The SDTF will be evaluating drift in canopy versus no-canopy situations.

4.4.2.3.5 Tank additives

The physical/chemical properties of the spray mixture or tank mix can have an important effect on the droplet size spectrum and thus on spray drift. Properties of known importance include viscosity, dynamic surface tension, and evaporation rate. The two mechanisms by which such an effect can occur are

1. directly influencing the size spectrum generated via atomization and
2. altering the evaporation rate of the atomized spray solution.

Viscosity is probably the most widely evaluated spray solution property and the factor most often modified in an attempt to control the formation of smaller droplets. "Thickening" agents or polymers have been added to spray tank solutions in an attempt to control droplet size formation. Such polymers have been shown to be effective in decreasing droplet break-up and thus increasing the volume mean diameter (VMD), although there is a limited amount of field work documenting whether there are actually decreases in drift associated with such tank additives. Additional work is required before quantitative predictions can be made on the percent reductions that can be expected.

After atomization, an important factor determining the amount of off-site movement is the evaporation rate of the smaller droplets. The addition of modifiers or surfactants may help reduce such evaporation. Surface active agents which reduce the surface tension and thus increase spreading can also reduce the evaporation potential. Other water soluble additives may also suppress evaporation. There are limited field data quantifying reductions in drift associated with products which reduce evaporation potential.

If tank additives that either increase the droplet size or reduce the evaporation rate are shown in comparative field studies to be effective in reducing drift, the additives offer a potentially effective mitigation approach. Product labels could require that specific "drift-reduction products" be used as tank additives. Such requirements would be enforceable. At this point, the limitation is data that demonstrate the effectiveness of such products.

4.4.2.4 Airblast applications

Application of chemicals to tree crops via airblast equipment represents the most complicated, and possibly most difficult to control, form of agricultural spraying. Differences in the geometry (height, width, etc.) and foliage density between trees make optimization of coverage extremely difficult. There is minimal flexibility in current application equipment, which severely limits the applicator's ability to adjust practices to varying conditions. This limited flexibility makes standardization of drift control practices difficult. Also, trees are fixed in their orientation and cannot be rotated for other crops as a procedure for pest control (as in row crops), thus minimizing flexibility in addressing pest control or drift mitigation. Because of these complexities and the need to maximize coverage, application practices that can be conducive to off-site movement are often employed.

Many of the concepts discussed for aerial drift relate directly to airblast spraying. However, there is an array of other factors that must be considered given the type of equipment and the importance of the target crop or canopy in determining drift potential from tree crops. In addition, there are unique aspects associated with different tree crops such that drift control measures may not be equally effective for all crops. For example, the height of pecan trees makes it impractical to apply many of the equipment-specific recommendations which may be applicable to apple or citrus trees. In addition, there are regional differences in application procedures and equipment that should be considered when evaluating mitigation options for tree crops.

Factors that influence target site deposition and product efficacy are not nearly as well understood for tree crops as they are for row crops. Efficacy is often inversely related to droplet size (efficacy increases as droplet size decreases). However, attempts to increase efficacy by reducing drop size can result in greater drift potential. When implementing drift reduction strategies by attempting to maximize the formation of larger droplets less susceptible to drift, care is required to ensure that product efficacy is not affected. Greater research is needed in application technologies that can increase deposition/coverage, which by definition should result in lower chemical usage while minimizing drift potential during the application process.

An important mechanism available for reducing drift for tree crops is associated with the development of new, specialized application equipment. Although theoretically this equipment can offer a short return time on investment via reduced chemical usage, such a return is primarily available for large production operations, where equipment is used on a more continuous basis. Such equipment is expensive and often not affordable for the smaller farmer, who represents an important segment of the tree crop industry. This factor should be considered when evaluating mitigation options for specific crops.

In comparison to aerial application practices, little quantitative information is available which allows predictions of the actual reductions in drift from tree crops associated with various mitigation options. To address this issue, the SDTF sponsored a workshop, organized by Dr. F. Hall of Ohio State University, to evaluate the options for reducing off-site movement of agrochemicals from tree crops. Below is a brief overview of some of the mitigation options discussed in this workshop. The actual report (SDTF, 1994) from the workshop should be consulted for more detail. In the workshop, three groups were formed to address three primary tree crop groupings: citrus, nuts, and fruits. Since many of the recommendations for mitigation were similar, this report will generically discuss all mitigation options, keeping in mind that direct applicability of any mitigation option will be dependent on regional and crop-specific requirements.

4.4.2.4.1 Application practices

Adjustments in the procedures used to evaluate the need for agrochemical applications and the specific procedures for making such applications represent a very practical approach for minimizing off-site movement. Lack of attention to specific application practices can actually lead to an increase in the number of applications required for pest control. It may be difficult, if not impossible, to actually quantify the reductions in drift which may be associated with some of the factors discussed below; however, all of these factors can be considered "directionally correct" relative to drift reduction.

Integrated pest management. Applying agrochemicals only under circumstances where they are required to maintain pest pressures below economic thresholds represents sound and well-established agricultural practices. Encouragement of such practices should reduce total pesticide usage, thus decreasing loading to off-site sites via drift.

Equipment calibration. Properly calibrating application equipment will ensure that excess chemical, beyond what is intended or required, is not being applied. Equipment should be calibrated and checked on a regular basis. Equipment checks include, but are not limited to, the measurement of applicator speed, nozzle output, and nozzle spray pattern. Equipment calibration could be considered mandatory for all application equipment, although quantitative data documenting drift reduction attributable to calibration are likely not practical.

Matching sprayer set-up with tree characteristics. The specific geometry and density of the target tree will determine the most appropriate sprayer set-up. Optimizing sprayer set-up for specific types of trees will increase coverage and reduce drift. For example, to maximize coverage and minimize drift for large trees with a low density of vegetation, a different gallonage, nozzle type, and configuration would be required than for a small, dense tree. Even within the same tree type, moving from one orchard block to another where the age of the trees is different may require a modification of the set-up. Often applicators will set up to ensure coverage of the largest trees and will not modify the set-up when they move to smaller trees. This can result in a portion of the spray applied to the smaller trees either missing or blowing through the smaller trees. Matching of the sprayer set-up with tree characteristics should reduce the amount of off-site movement that can occur when such matching is not performed. Unfortunately, the inflexibility of many of the current sprayers makes such matching difficult to perform in the field, thus contributing to the lack of matching under many circumstances. Currently, data do not exist on the reductions in drift which could be attributable to matching versus not matching in equipment set-up.

Expression of application rate. As alluded to in the discussion on matching sprayer set-up to tree size, greater understanding in calculating and expressing application rates is needed. There is great diversity in the size and density of fruit trees. A standard per-acre rate cannot account for such differences. The most appropriate method is to calculate rates based on tree (orchard) size (often referred to as tree-row-volume). Considerable information is available on this topic, and numerous labels have methods for determining accurate size-dependent rates. There is currently a need to standardize these labels and educate growers/applicators on the importance of this concept in minimizing pesticide use.

Application timing. If possible, the timing of applications should coincide with periods or conditions which are least conducive to drift. Such timing generally refers to making applications during periods of the day when weather conditions are optimal for on-target deposition. Seasonal timing is generally not applicable since the applications are controlled by pest pressures, as identified by IPM programs. Wind is normally minimal during early morning or late afternoon and into the night. However, since this is also the time when temperature inversions are most likely to occur, applications at these times cannot always be associated with lower drift potential. Generally, a low wind (3 to 4 mph) is best for application because the directional movement of the droplets is predictable, and drift distance should be minimal. Application timing restrictions should be considered only on a regional and crop-specific basis and must be evaluated based on the specific circumstances of the product being evaluated. Overall, timing is primarily a weather issue and may be best handled with weather restrictions.

One-way spraying. Requiring that applications to trees in the outer 1 to 5 rows be directionally restricted to spraying only toward the inside of the orchard represents a potentially useful approach for minimizing drift. Some estimates indicate that the complete spraying (*i.e.*, both sides of the tree) of the outer 4 to 5 rows accounts for greater than 90% of the off-site drift. By spraying only one side of the trees at the inward direction, the amount of drift may be reduced. Presently, data do not exist that indicate such spraying will reduce drift. For dormant trees with no vegetation, it is doubtful that such spraying will reduce drift; however, for full canopy trees a reduction may be observed. A problem with such an approach is the potential for decreased pest control in the outer rows due to decreased coverage.

Farm management plans. Within any given orchard, it is likely that only a portion of the orchard is adjacent to sensitive areas of concern which will require mitigation to reduce drift. Given this fact, it may not be practical or necessary to apply the specific mitigation measures to the entire orchard. Sensitive or critical blocks within the orchard can be identified, and specific drift control or management plans developed to minimize or reduce drift from those specific blocks. The management plan could be specific for each farm and draw on all the available options for minimizing off-site movement. Different growers, given different conditions or situations, could develop different approaches for reducing drift. To use a specific product which exceeds an LOC, a requirement could be imposed for the grower to develop (and possibly submit to the state) a management plan for reducing drift.

Tank additives. The tank-additive concepts discussed in aerial applications are also applicable to tree crops, although no data are available documenting that additives can be used to reduce off-site movement. Concerns exist regarding the influence of tank additives on product efficacy and fruit finish/quality. Until field data are available to document reductions, tank additives are not considered a viable drift reduction option for airblast operations.

4.4.2.4.2 Equipment

Equipment represents an important variable which can influence off-site movement. The majority of applications to tree crops are made using axial flow airblast sprayers equipped with hollow cone hydraulic nozzles. Although these sprayers can be set up for specific tree crops, the overall flexibility of the sprayers to optimize coverage and minimize drift is low. New, more flexible equipment is available; however, such equipment is expensive, and limited data are available on actual reductions in drift which can be attained by such sprayers.

When considering equipment, either the type of sprayer or the nozzle used on the sprayer can influence off-site movement. The set-up of the sprayer can also influence drift, and this was discussed in the previous section (4.4.2.4.1 Application Practices).

Nozzle selection. Nozzle type influences droplet size, which will directly influence off-site movement. However, the complexity of tree crop spraying makes it difficult to generalize about nozzle selection. For airblast sprayers, there are different nozzle sizes located at different positions on the sprayer. Often multiple products applied in the same tank mix have different needs concerning coverage, and changes in the droplet spectrum may influence the efficacy of the different products in different ways. In general, data are lacking on the actual droplet distributions associated with different nozzle configurations used in airblast sprayers, although the SDTF is beginning to develop some data on the influence of nozzles and pressure on droplet size.

A general guide to nozzle selection needs to be developed. Such a guide should address nozzle selection for maximizing coverage (smaller droplets) and minimizing drift (larger droplets). Presently, such standardized guidance is not available.

Application equipment design. Although the large majority of equipment used in tree crops are axial flow airblast sprayers, other commercially available types of equipment theoretically should result in less off-site movement. In many cases, the equipment was developed to increase application efficiency, and thus increase efficacy and decrease chemical usage. By default, this should also result in decreased off-site drift. Conversion to such equipment represents a long-term improvement in application technology; however, conversion is not an immediate solution. The equipment is expensive and will only slowly replace the older, less efficient, and inflexible models.

Currently, very little data exist on the relative reductions in drift which may be attributable to the newer equipment. Specific types of application equipment that should be considered include:

- Sensor-type sprayers—External sensors are used to identify the presence or absence of a canopy. Spray output is controlled, in an on-off fashion, relative to the presence or absence of vegetation (individual trees). When there are open spaces between trees, the sprayer automatically shuts off, thus avoiding spraying into open air. This reduces chemical use and may reduce drift.
- Tower sprayers—A variety of configurations of tower sprayers exists. In general, a funnel is used to vertically transport the air flow up and down over the top of the canopy. Nozzles can be positioned at various locations within the air stream. The spray solution is thus directed downward into the tree canopy, as opposed to upward through the canopy. Drift should be reduced because the transport of droplets up and out of the canopy is minimized. Due to tree size and orchard topography, such sprayers may not be applicable to all types of tree crops.
- Cross-flow sprayers—Air flow from these sprayers is horizontal, and the chemical is sprayed directly into the canopy. Individual fan units are stacked vertically up to the height of the trees, with each unit containing the appropriate nozzles. Since all air flow is horizontal, movement of droplets above the canopy should be minimized.
- Shield or tunnel sprayers—The tree is completely enclosed and sprayed, from all sides, via nozzles within the tunnel or shield. Such sprayers are limited in practical use to smaller trees.

Growers are encouraged to switch to such equipment; however, until the equipment becomes more common and data are available on the reductions in drift associated with specific types of equipment, the use of these sprayers should not be considered as a mandatory drift reduction requirement.

4.4.2.4.3 Meteorology

Wind speed, wind direction, and air stability can all influence off-site movement. The rationale for such restrictions is the same for tree crops as for row crops, and such restrictions should be considered when trying to reduce off-site movement. Unfortunately, quantitative data do not exist which will allow predictions of the reductions in drift that may be attributable to meteorological restrictions. Data from field crops will not be directly applicable to tree crops, given the influence the tree canopy can have on off-site movement. The SDTF hopes to provide data that will allow predictions of the influence of meteorological variables on drift.

4.4.2.4.4 Site modifications

Physical modifications of the site which may reduce off-site movement include the use of buffer zones and, in the long term, physical wind breaks and the conversion to smaller trees.

Buffer zones. The majority of the factors on buffer zones discussed in the general drift section are applicable to tree crops. Buffer zones can directly reduce the amount of product which moves off-site; however, at this point, limited quantitative data exist to predict what reductions may be associated with specific distances. The SDTF is currently generating data that are expected to provide such predictive information. Relative to the use of buffer zones, specific factors which must be considered include the following:

- Tree crops are semi-permanent, and there is limited flexibility associated with establishing a buffer zone.
- There is a high cost associated with taking trees out of production. In new orchards, "wind break" trees could be planted in the buffer area.
- There are specific pros and cons associated with leaving trees within the buffer zone versus cutting them down.

  Pros

  - Trees left in the buffer area can be treated with other products to control the target pest, assuming alternatives are available which do not contain the same buffer restrictions.
  - If the label has wind direction specifications, then trees in the buffer area can be treated as long as wind is not toward the area of concern.
  - Trees in the buffer area can provide an effective filter which removes droplets that may be drifting from applications within the orchard.

  Cons

  - If trees within the buffer area are not treated, they can act as a major source of pest re-infestation. This fact may result in the increased use of chemicals for pest or disease control.

Relative to row crops, buffer zones in tree crops are more complicated to implement and can have greater direct impact on the grower. However, at this point, buffer zones represent one of the few mitigation options that will have a direct, quantifiable reduction in off-site movement.

Wind breaks. Off-site movement can be influenced by erecting structures, either natural or manmade, between the sensitive off-site areas and the tree crops. Most often, the planting of coniferous trees is discussed as a viable wind break option, although the use of artificial wind breaks has also been mentioned. Either option represents medium- to long-term options for reducing drift.

To be effective, such trees must allow air flow to ensure the filtering of the moving droplets. If the vegetation is too dense, the wind break may increase the vertical movement of droplets over the trees and higher into the air flow. In addition, if the wind break restricts air movement, frost and/or increased disease problems may occur within the orchard.

Presently, very limited data exist on the effectiveness of such barriers in actually reducing drift. Until data that establish the effectiveness of wind breaks are developed, wind breaks should not be considered as a primary option for drift mitigation.

Conversion to smaller trees. Application to smaller trees offers more opportunity for controlling off-site movement. For production reasons, the trend in tree crops is toward the planting of smaller trees. If appropriate for the crop, this trend should be encouraged and should, over time, provide the type of canopy which will allow for greater control of off-site drift.

4.5 MITIGATION RECOMMENDATIONS

- The cluster concept should be applied to mitigation when appropriate. Clustering can be based on crop groups and/or specific pesticide properties.
- Mitigation must be cost effective and/or cost responsible, and mitigation implementation must be practical (with consideration to the agriculture community's experience).
- Successful mitigation compliance may necessitate educational programs for the user.

REFERENCES

Baker, J.L. 1983. Agricultural areas as nonpoint sources of pollution. In M.R. Overcash and J.M. Davidson, eds., *Environmental Impact of Nonpoint Source Pollution.* Ann Arbor Science Publishers, Ann Arbor, MI, pp. 275-210.

Baker, J.L. 1987. Hydrologic effects of conservation tillage and their importance relative to water quality. In T.J. Logan, J.M. Davidson, J.L. Baker, and M.R. Overcash, eds., *Effects of Conservation Tillage on Ground Water Quality,* Chapter 6. Lewis Publishers, Chelsea, MI.

Baker, J.L. and J.M. Laflen. 1979. Runoff losses of surface-applied herbicides as affected by wheel tracks and incorporation. *J. Environ. Qual.* 8:602-607.

Baker, J.L. and J.M. Laflen. 1983. Water quality consequences of conservation tillage. *J. Soil and Water Conservation.* 38:186-193.

Bovey, R.W., C. Richardson, B. Burnett, M.G. Merkle and R.E. Meyer. 1978. Loss of spray and pelleted picloram in surface runoff water. *J. Environ. Qual.* 7:178-180.

Dillaha, T.A., R.B. Reneau, S. Mostaghimi and D. Lee. 1989. Vegetative filter strips for agricultural nonpoint source pollution control. *Trans. ASAE* 32:513-519.

Felsot, A.S., J.K. Mitchell and A.L. Kenimer. 1990. Assessment of management practices for reducing pesticide runoff from sloping cropland in Illinois. *J. Environ. Qual.* 19:539-545.

Glenn, S. and J.S. Angle. 1987. Atrazine and simazine in runoff from conventional and no-till corn watersheds. *Agric. Ecosystems and Environ.* 18:273-280.

Hall, F.R. 1994. Final report of the work group on tree crops drift mitigation. Arlington, VA, September 14-15, 1993. SDTF, Washington, DC.

Hall, J.K., N.L. Hartwig and L.K. Hoffman. 1983. Application mode and alternative cropping effects on atrazine losses from a hillside. *J. Environ. Qual.* 12:336-340.

Hall, J.K., M. Pawlus and E.R. Higgins. 1972. Losses of atrazine in runoff water and soil sediment. *J. Environ. Qual.* 1:172-176.

Martin, C.D., J.L. Baker, D.C. Erbach and H.P. Johnson. 1978. Washoff of herbicides applied to corn residue. *Trans. ASAE* 21:1164-1168.

Rhode, W.A., L.E. Asmussen, E.W. Hauser, R.D. Wauchope and H.D. Allison. 1980. Trifluralin movement in runoff from a small agricultural watershed. *J. Environ. Qual.* 9:37-42.

Triplet, G.B., Jr., B.J. Conner and W.M. Edwards. 1978. Transport of atrazine and simazine in runoff from conventional and no-tillage corn. *J. Environ. Qual.* 7:77-83.

Wauchope, R.D. 1978. The pesticide content of surface water draining from agricultural fields - a review. *J. Environ. Qual.* 7(4):459-472.

USEPA. 1992. Guidance specifying management measures for sources of nonpoint pollution in coastal waters. Washington, DC.

Chapter 5: VERIFICATION OF MITIGATION EFFECTIVENESS

The purpose of this chapter is to address approaches for the verification of the effectiveness of the mitigation steps outlined in Chapter 4.

Understanding the causes of decline of natural resources can be problematic. Neither the U.S. Environmental Protection Agency's (USEPA) Office of Pesticide Programs (OPP) nor the registrant community has the resources to create a field program adequate for long-term, broad-scale investigation of the effects of pesticide use on the ecosystem. Verification of the efficacy of pesticide risk mitigation measures must, therefore, be undertaken with careful attention to the allocation of human and fiscal resources so as to obtain the maximum reduction of total pesticide risk.

5.1 PHILOSOPHY FOR VERIFICATION

Although it is believed that mitigation measures can reduce environmental exposure and thus potential adverse effects, there has been virtually no systematic identification or evaluation of appropriate mitigation measures for pesticides that exceed ecological levels of concern (LOC). The current efforts of the Spray Drift Task Force (SDTF) represent the most thorough evaluation to date (relative to drift mitigation), although the primary objective of the SDTF was not the evaluation of mitigation effectiveness. Because of the difficulty and cost of demonstrating cause and effect relations, verification approaches must be practical and sound. USEPA's approach to accomplishing these goals should be to utilize a variety of verification activities in a stepwise fashion. In the array of mitigation actions that might be proposed for a pesticide or group of pesticides, some of those proposed actions will be based on best professional judgments (reduction of application rate); some will be supported by scientific data (buffer strips, ratio of spray boom length to wing span, etc.); and some will be unsupported by scientific data but reasonable or "directionally correct." In the identification of the risk, there may be missing scientific data needed to clarify an issue or the extent of risk. **Verification** in the context of this paper is the evaluation and/or collection of data needed to address pivotal issues in risk reduction. In considering verification steps, it is necessary to recognize the dynamic side of pesticide registration (data are not available to address the emerging issues and must be developed). The **new paradigm** attempts to put into policy a means to protect the environment while developing and building data bases that continuously allow for improved regulatory decisions. Verification of the risk reduction will depend on the proposed mitigation measure or the information that is required to clarify the extent of the risk. The policies implemented with the new paradigm deal with immediate regulatory requirements for pesticides and also direct USEPA to develop methods for assessing long-term effects.

Biological monitoring can play an essential role in evaluating the status of ecosystems and the impacts of pesticides. When verification is required for evaluating mitigation designed to address potential long-term effects, biological monitoring can be used in conjunction with chemical monitoring, although it should utilize existing state and federal surveillance programs whenever appropriate.

For mitigation verification, the USEPA should attempt to utilize existing federal programs that are evaluating the condition of the nation's water resources. Utilization of such programs could come about through cooperative research agreements, direct monetary support of ongoing projects, joint industry and USEPA ventures under the Federal Technology Transfer Act, and through better utilization of research findings by state and federal research institutions and state-funded surveys. In order to ensure that regulatory measures are correct and protective, the USEPA should continue its effort to coordinate and cooperate with other federal agencies to obtain and develop the data bases necessary to make the appropriate decisions regarding the protection of the nation's natural resources.

5.2 ASSESSING AND MONITORING THE SUCCESS OF RISK MITIGATION MEASURES

Verifying the success of mitigation measures in reducing risk can be thought of as a series of steps of increasing complexity and cost, steps to be taken only if the risk assessment process so requires. If the initial risk assessment concludes that risk is either minimal or excessive, the mitigation verification process need never be undertaken. In the intermediate cases, mitigation will be evaluated and verification may be needed.

Risk—the probability or frequency of harm—is an amalgam of the irreducible natural variation governing the frequency of occurrence of significant exposure events and toxicologically sensitive species and the scientific uncertainties that distort estimates of exposure and hazard. Natural variation generally cannot be controlled or regulated: It is an intrinsic risk in agricultural production or any other human activity. Thus, the mitigation verification process will be driven in large measure by the uncertainty component of the risk assessment; this process ends when risk can be reliably characterized.

The final regulatory endpoints may take several forms. Risk may be excessive and uncontrollable—leading to withdrawal of the chemical. Risk may be controlled by mitigation to tolerable levels—leading to relaxation of study requirements or other restrictions on the use of the chemical. In either case, the verification process will have reached closure by resolving the scientific uncertainties that required clarification.

Although the focus of mitigation is on preventing harm to non-site natural resources and ecosystems, the verification program must be designed around the realities of the costs of *per se* natural resource monitoring and the large natural variability of wildlife populations and ecosystems. The verification program must maximize human and fiscal resources as well as the reduction of total pesticide risk to sensitive populations and systems. This verification can be accomplished through a series of increasingly complex investigations, coupled with continual incorporation of the new data to refine and clarify the existing risk assessment.

When a pesticide has raised concern about potential ecological damage, proposed mitigation measures will have been negotiated between the registrant and the USEPA. Given these agreed-upon measures, it will be important to verify that the original concern has in fact been alleviated, *i.e.*, risk has been reliably reduced to acceptable levels. Verification will focus resources on the most significant sources of uncertainty in this evaluation. Three potential "phases" are envisioned:

1. Verification that the proposed mitigation is adequately responsive to the ecological concerns—This initial judgment may be entirely qualitative or may include a probabilistic quantification, in some cases using the full risk assessment process. In any case, this phase will be conducted by using existing data sets; it will not usually involve additional data development.
2. Field studies to improve estimates of the success of mitigation—These may include surveys to evaluate the ability of the user community to implement mitigation measures in the field, laboratory experimentation to refine estimates of model parameters that are specific to mitigation processes, and field experiments to refine estimates of mitigation efficacy and uncertainty in mitigation practices under the use conditions and crop/region of concern.
3. Field investigations of the actual achieved performance of mitigation, and wider scale analysis of environmental residues of the pesticide—These may include sampling of environmental residues in entire watersheds, sampling of tissue residues or biomarkers of exposure. Under compelling circumstances (*e.g.*, recurrent incidents, documented degradation), study of the status and dynamics of the specific organisms and ecosystems of concern may be necessary.

This description of a "phased approach" should not be taken to mean that it is necessary to start at the first phase and proceed through subsequent stages. Entry into the system will be based in part on the factors that triggered the need for mitigation and on the status within the risk assessment process. For example, if individual edge-of-field models addressing the specific mitigation options are fully validated, uncertainties in the risk characterization cannot practically be reduced, and yet risk estimates are unsatisfactory, an immediate move to Phase 3 (surface water residue monitoring, for example) may be appropriate. Logical, timely, and environmentally protective decisions that allow further action to proceed can be taken at any point during the verification process; however, it is not a process that must run its allotted course under all circumstances.

Validated mathematical models and computer-based management systems will be critical elements of this program because they will see continual use as a formal evaluation mechanism in every tier of the process. Modeling will be used to estimate exposure and thus identify ecological concerns posed by specific pesticides. These same models can be used to assess the worth of the suite of proposed mitigation measures and to evaluate the success of mitigation as realized under field conditions. The models combine basic physical, chemical, and biological principles with the specifics of an ecosystem biogeography in growing areas. The empirical data bases—known drainage properties of cultivated soils, potential wind and water erosion under a variety of tillage systems, etc.—used to parameterize the models are critical to conducting accurate assessments. These data will require continual maintenance and update as monitoring programs and appropriate research programs acquire results. The process of creation and maintenance of a computerized data base of known quality and applicability will create a formal "institutional memory" that will benefit both food production efficiency and natural habitat in the application sites.

Information related to the patterns of use or acceptance of mitigatory steps by the user community is important and can be obtained by utilizing existing information-gathering systems within the federal sector, through grants to state universities and colleges, or via the record-keeping provisions of "restricted use" classification. The availability of existing infrastructures to collect and collate information that could be used to verify the effectiveness of mitigation should be considered. "Restricted use" classification could be considered as one of the first mitigation actions for pesticides that raise an aquatic risk concern. The results of such efforts can potentially be used in exposure modeling programs to evaluate whether ecological concerns have been alleviated under field conditions.

The "restricted use" classification would make a number of tools available for characterizing the success of mitigation and provide a mechanism for monitoring the actual use of the product. "Restricted use" products are those whose use poses a unique risk to man or the environment. Specific requirements under the Federal Insecticide, Fungicide and Rodenticide Act (FIFRA) for application records apply to all "restricted use" products applied by either commercial or private certified applicators. Required records are available for inspection by appropriate federal and state control officials on an as-needed basis. Such records could be an important element of specific programs to assist in the overall appraisal of the adequacy of mitigation actions.

5.2.1 Phase 1

Phase 1 of verification is devoted to formal inference methods (statistical inference and simulation modeling) and expert judgment, using available data. If risks can be demonstrated to be mitigated to low levels with high certainty, then mitigation will have been "verified" by definition, and no additional action may be required. The reliability of this process flows from its use of the very same inference tools that have been used to infer the existence of significant risk and initially set in motion the mitigation process.

Given a set of appropriate mitigation proposals, three important activities can be identified:

1. use of simulation models to establish effectiveness,
2. use of best professional judgment and statistical inference in evaluating mitigation effectiveness, and
3. evaluation of acceptability and practicality in the user community.

A variety of data bases and simulation models are available for evaluating the effectiveness of specific mitigation practices. For mitigation methods addressing spray drift, the final models to be validated by the SDTF will have the capability of specifically addressing exposure reductions expected from the suite of practices. For runoff, Table 5-1 outlines the capabilities of currently available models.

TABLE 5-1. Basic Pesticide Capabilities of Primary/Secondary Models

	EPIC	GLEAMS	PRZM2	SWRRB	EXAMS	WASP5	DESPM	AGDISP
Field-Scale	x	x	x					
Watershed-Scale				x				
Surface (Edge of Field)	x	x	x	x				
Rootzone (Bottom of Rootzone)	x	x	x	x				
Subsurface Flow (Edge of Field)	x							
In Water Body Concentration					x	x		
Daily Time Step	x	x	x	x	x	x		
Less Than Daily Time Step					x	x		
Multiple Pesticides	10	10	3	10				
Formulations								
Application Timing	x	x	x	x				
Application Rates	x	x	x	x				
Crop Yields Simulated	x							
Strip Cropping								
Crop Rotations	x	x	x	x				
Irrigation	x	x	x	x				
Surface Drainage								
Erosion	x	x	x	x				
Terraces	x	x	x	x				
Detention Ponds (Fate)				x	x	x		
Streams (Fate)					x	x		
Lakes (Fate)					x	x		
Buffer Strips		x						
Set Backs		x						
Grassed Waterways	x	x		x				
Incorporation	x	x	x	x				
Banding								
No-Till	x	x	x	x				
Conservation Tillage	x	x	x	x				

In general, these models need further development to routinely accommodate mitigation options and evaluate them in context. The ultimate goal is the validation of a suite of models, firmly grounded in physical and chemical processes, that reliably represent the capture, storage, and degradation of chemicals. Although the algorithms for these models are for the most part already known, specific additions to the larger models remain to be accomplished. In addition, the data base of model-specific parameters and their distributions must be assembled so that they can be used in a probabilistic risk assessment context.

Data bases that provide information on current agricultural practices and statistics are also available. The U.S. Department of Agriculture (USDA) Soil Conservation Service (SCS) utilizes the CAMPS system and is in the process of converting district offices to the Field Office Computing System (FOCS). Official practices are recorded by field, farm, and producer for all participants in SCS programs such as the Food Security Act (FSA). USDA-Agricultural Stabilization and Conservation Service (ASCS) maintains a separate data base for cost-sharing participants. This information has commonalities with the SCS data but is not coordinated between the two agencies. Record keeping for all "restricted use" pesticides, with details of rates and methods of application, became law in May 1993 and may provide valuable information for evaluating mitigation options.

Information on cropping practices, pesticide use, etc., can be valuable in evaluating mitigation options. As an example, these data bases can be queried for the percentage of soybean acres on a particular soil with a slope greater than 6% in a particular watershed that is using buffer strips during a particular year. This indicates, given the methodological efficacy of buffer strips, the maximum achievable mitigation.

A preliminary assessment of some of the range of efficacies of mitigation options was provided in Chapter 4. Intended as a starting point, these estimates represent an initial best judgment of the benefit expected from each practice, operating alone. These efficacies are not always simply additive. Residues that are mitigated by one method may detract from the efficacy of another (by competing for the molecule), may enhance the efficacy of another (by directing the molecule along an appropriate transport path), or may be independently additive. Only the spatially explicit simulation model, based on mass balance, site geometry, and mitigation process, will be able to evaluate these interactions with any precision. However, for the moment, best professional judgment will have to suffice.

The tentative values and large uncertainties associated with mitigation efficacies point out the need for a focused effort to develop and validate the specific algorithms, parameters sets and distributions, and statistical experience needed to characterize the expected value and uncertainty of mitigation options under the entire suite of geophysical, climatic, and pedological conditions found in American agriculture. The creation of this data base is a high priority recommendation.

In some cases, logical evaluations may suffice for verification. For example, the mitigation effectiveness of application practices (banding, for example) is apparent and needs little further evaluation. When lower limits of single practices suffice to reduce risk to acceptable levels, the risk may well be concluded to be acceptably mitigated.

5.2.2 Phase 2

At the time of proposal, the field practicality of mitigation measures will also be relatively uncertain. The modeling used to assess the plausibility of mitigation will make use of prescribed nominal values for both the reduction of off-site transport of the pesticide and for the acceptance and field application success of mitigation measures in the affected community. Clearly, however, both these elements may require some degree of certification through field investigations. A concerted program of field validation of the physical efficiency of mitigation measures is, again, strongly recommended, and this program may achieve an adequate characterization of the expected values and variance of pesticide risk reduction as a function of pedology, topography, weather, etc. The matter of practicality (*i.e.*, the ability of the user community to apply the mitigation measures consistently and effectively) remains.

When ecological concerns have not been alleviated by evaluation of the proposed mitigation methods in Phase 1, then a number of courses of action, as alternatives to immediate cancellation, may be undertaken. These include the following:

- Proposal of additional mitigation measures and return to Phase 1 risk assessments if there is confidence that the first measures have proven inadequate—The identification of a need for additional mitigation measures or alterations to correct control problems should always be an open option at any phase of the process.
- Efforts to achieve greater field acceptance of mitigation if data indicate that users are simply not complying with the measures—This may involve improved education programs and follow-up surveys to evaluate changes in behavior.
- Studies better quantifying the risk reduction achieved by each individual mitigation measure under the specific use conditions—Such studies are addressed in greater detail below.
- Site-specific investigations of the combined efficacy of the currently achieved practice—Such studies are addressed in greater detail later.

Because all mitigation measures may not be 100% practical, a measure should not be accepted if it requires so stringent an application that there is no room for honest error, natural misfortune, or industrial accident. Verification of the acceptance and practicality of mitigation measures can clearly be accomplished only through strong cooperation among the USEPA, the farming community, and USDA field programs. An important aspect of verification will be an examination of the realized experience of the agricultural community in implementing the proposed mitigation measures. User surveys may be the most appropriate mechanism for obtaining this information.

Feedback from the user community is essential and will serve several purposes. First, the user community will be made aware of its success rate in applying mitigation measures, for its own guidance and for evaluation of the feasibility of retaining the product as a pest control tool. In many instances, use surveys will be undertaken in circumstances under which mitigation is failing, and if measures are indeed being fully implemented, withdrawal of the chemical may be necessary. Field surveys will help establish whether additional education and training may help to achieve full implementation of mitigation and its expected benefits.

When modeling is not adequate to evaluate mitigation (Phase 1), then experimental field trials may be required to investigate specific mitigation measures. In such an evaluation, comparative studies simulating field conditions may be required to develop predictive capabilities of models. Measuring off-site transport can be linked with modeling to determine the effect of specific mitigation measures and to allow a prediction of mitigation effectiveness. Verification/calibration of models is essential in order to improve the predictive power of models or to allow extrapolation of the study results to other sites. In general, validation is generic and is related to specific pesticides by chemical parameters. No model can be validated for every compound in every soil/climate/management scenario.

The principal routes of exposure that must be modeled are spray drift and runoff. The SDTF data base should provide an adequate basis for predicting the effects of mitigation. The landscape factors that may affect exposure, such as wind speed, can be included in exposure models for evaluating mitigation or for refining the risk assessment. Mitigation of exposure by minimizing spray drift will be dependent generally on application practices rather than compound, and the models developed by the SDTF will provide predictive information that will be applicable to most pesticides. The models should be readily available, and exposure reduction factors should be widely accepted.

Evaluating specific measures to reduce runoff losses and delivery of pesticides can be done best through small-scale, comparative studies that examine factors such as buffer zones, application methods, formulations, or conservation practices. Studies simulate field conditions using well-characterized plots in which water, soil, and chemical transport are controlled through rainfall timing, intensity, and duration. The studies lend themselves to model validation as well as development of reduction factors for minimizing transport off-site and delivery to a surface water resource.

Specific advantages and disadvantages of small-plot experimental runoff studies are listed below.

Advantages

- Methods are available.
- Predictive information is obtained.
- Model validation is possible.
- Important variables can be controlled.
- Site characteristics can be controlled.
- Site selection is simpler than for large watershed studies.
- Small-plot experimental runoff studies are cost efficient relative to large watershed studies.
- Statistical evaluation of specific variables can be obtained.

Disadvantages

- Scaling to field size is critical.
- Concentrations may represent an extreme case and must be properly interpreted.

Site-specific studies can be used to evaluate the combined effect of all mitigation factors. In site-specific studies, the object is to study the fate of the pesticide where the application is controlled and where the off-site transport is measured. Runoff and drift will occur as a natural result of the conditions at the site, and the fate of the pesticide on the site or in the receiving system will be investigated under the measured conditions. Model validation information can be obtained from such studies; extensive characterization of the site is required, however. The studies are not as amenable to model validation work as the small-plot studies.

Specific advantages and disadvantages of site-specific studies follow.

Advantages

- Source (drift, runoff) is directly related to exposure.
- Scaling concerns are obviated since the study is conducted in a field.
- Application can be monitored and controlled to ensure that mitigation measures are followed.
- Biological effects may be studied at the same time and linked to exposure.
- Study conditions reflect the real world at that site.
- Runoff, drift, and/or fate models can be validated at a completely characterized site.

Disadvantages

- Replication is difficult.
- Predictive value is low unless models successfully describe the results.
- Site characterization may be costly.
- Results are dependent on weather and site conditions.
- Key variables cannot be controlled.
- Site selection is difficult.
- Model validation is difficult.
- Results are applicable to only one site, which may be atypical.
- Results cannot be extrapolated to other sites.

The remainder of this section provides an example of how one might proceed through Phases 1 and 2, and possibly into Phase 3. A generic situation is provided which illustrates some of the rationale and decision-making steps.

Phases 1 and 2 Verification Example. Based on the risk characterization, an insecticide used on a major agricultural crop has exceeded a level of concern (LOC) for acute and chronic effects on aquatic invertebrates. No incident data are available, although the predictive risk assessment indicates impacts on invertebrates may result in secondary ecosystem effects. The product is applied four times per year on a 14-day interval by ground or air and is used throughout the Midwest. Transport to the aquatic environment can occur by either runoff or spray drift; however, drift is considered the primary exposure route of concern. The compound is highly hydrophobic such that surface runoff loss, when it occurs, will primarily be associated with soil erosion.

Based on exceeding the LOC, mitigation options agreed upon between the registrant and the USEPA include

- 150-foot buffer areas to reduce off-site movement from drift;
- 25-foot vegetative filter area to reduce erosional transport; and
- on highly erodible lands, the requirement to follow SCS recommendations for erosion control, as stipulated in the 1985 Food Security Act.

These mitigation options were chosen based on best professional judgment, and there are limited data quantifying the reductions in exposure that may be realized. There is the need to determine whether the mitigation options reduce risk to an acceptable level. (**Note**: Such a determination should entail evaluating the uncertainty in both the exposure and hazard components of the risk assessment. If there is a need to improve the predictive nature of the hazard component, then additional effects data, as discussed in Section 2.2 of Chapter 2, should be requested. In the present situation, given the hydrophobic nature of the product, it may be necessary to address uncertainties associated with toxicity of the product to sediment dwelling organisms.)

Relative to evaluating the mitigation's exposure reduction effectiveness, the primary areas of uncertainty are these:

- How effective is the 150-foot buffer in reducing drift?
- How effective are the grassed buffer strips and general SCS-approved erosion control measures in reducing pesticide transport?

Given these uncertainties, the ability of currently available data to determine mitigation effectiveness should first be reviewed (Phase 1). For the spray drift evaluation, the data base being generated by the SDTF should be adequate to quantify the reductions in exposure associated with a 150-foot buffer zone. No additional data should be required to address this mitigation measure. For the runoff mitigation options, adequate quantitative data may not be available for predictive assessments. In this case, the initial verification phase would focus on developing data to predict the reductions in exposure associated with the specific mitigation options (Phase 2). Small-scale or site-specific runoff studies, with the primary focus being model validation and effectiveness of vegetative filter strips, may be appropriate.

When the spray drift assessment and runoff model validation work has been completed, the risk assessment should be revisited based on the new exposure predictions (incorporating the mitigation options) and any additional hazard information also requested for refining the hazard assessment. If the erosion control practices and buffer zones reduce exposure below the LOC based on the use of the validated model and standard risk assessment procedures, then additional work should not be required. If at this stage the LOCs are still exceeded, additional regulatory action may be necessary. This action could entail a number of options, including removal of the product from the market, the request for additional mitigation measures, or the request for additional data to further define the risk.

Assuming product cancellation is not an option, potential scenarios addressing the "next step" question include

- implementation of additional mitigation options and cycling through the Phase 2 approach again, in which focused studies are performed to generate predictive data (model validation); or
- if additional mitigation factors are not available, or if it is determined that it is not appropriate to implement additional factors, movement to Phase 3 verification.

Phase 3 would involve more detailed retrospective evaluation of the resource at risk and could include watershed residue monitoring and/or biological monitoring. The most appropriate approach should be based on the remaining uncertainty in the risk characterization. Watershed monitoring will provide data on actual residues in the receiving system and will integrate all of the mitigation and landscape factors controlling exposure. In the present example, additional factors likely not considered in the exposure assessment, such as the influence of riparian vegetation on drift and runoff and the probability of drift occurring during a particular application, are considered in a watershed approach. If available, exposure biomarkers could be useful to establish whether organisms are actually exposed to significant residues. Biological monitoring should be considered only after watershed residue monitoring indicates that the actual exposure still exceeds the LOCs.

5.2.3 Phase 3

Phase 3 includes field investigations of the actual achieved performance of mitigation and wider scale analysis of environmental residues of the pesticide. Such investigations may include statistical sampling of environmental residues in entire watersheds, sampling of tissue residues or biomarkers of exposure, and, although only under very unusual circumstances, study of the status and dynamics of the specific organisms and ecosystems of concern.

In unusual circumstances, the effect of a specific pesticide on a particular ecosystem will require residue or effects monitoring to confirm cause-effect relationships or to conduct a final confirmatory risk assessment. These studies will be required only when there is a high, but still uncertain, probability of an effect, when a biological effect has been demonstrated (through incident reports, for example), or when mitigation measures cannot be imposed without the loss of the pesticide benefits.

Exposure. General exposure surveillance monitoring may be appropriate when it is necessary to determine actual exposure to a widely used chemical within a watershed or on a large scale. Such studies should only be required when, after model validation, predicted high risks persist due to uncertainty in the realized performance of mitigation practices. Surface water and sediments may be monitored at specific locations within the watershed, but sampling will often not be linked to specific sites, applications, or mitigation measures. Such studies inherently include the influence of all landscape factors on exposure, and thus residue measurements represent actual transport to the aquatic environment. These studies must undergo a careful problem formulation and statistical sampling design phase.

The design phase of a watershed-scale monitoring study should include among its objectives the validation of the expected value of the suite of mitigation measures in use in the watershed, under field conditions. The problem formulation phase must include an analysis of the degree of change in the expected value and variance of mitigation effectiveness that will be accepted as sufficient to confirm or reject the current risk assessment. The statistical design must include a power analysis to determine the probability of Type I and Type II errors and, whenever possible, comparisons among control watersheds or among a variety of mitigation treatments.

Specific advantages and disadvantages of this approach include the following.

Advantages

- Environmental concentrations are measured.
- Adequate numbers of sites and samples can give exceedance curves.
- State and federal monitoring networks already exist.
- Transport downstream is measured.
- "Paired" watersheds can be used to conduct comparative studies.
- Integrating existing monitoring networks can leverage costs and resources.
- All modes of exposure (drift, runoff, direct overspray) are included.
- Landscape factors are included in measured concentrations.

Disadvantages

- Representative sampling schemes for sediment must be developed.
- Results cannot be extrapolated to other sites, except when careful model validation studies are linked to the study design.
- Predictive value is slight.
- Sources of pesticide are unknown.
- Effect of specific mitigation factors is unknown.
- Complete characterization of the product's effect is likely to be very costly.
- Interpretation requires information that will be difficult to obtain, such as total pounds applied or application timings.

Tissue residue analysis and chemical-specific biomarkers, such as measurement of depressions in acetylcholinesterase activity as an indicator of exposure to organophosphate insecticides, may be used in conjunction with residue monitoring. Such measurements can be indicators of actual exposure of the non-site organisms to the chemical of concern. Effective use of such a measurement requires establishing the dose-response relationship between the change in the biomarker of interest and exposure to the chemical. High variability requires adequate sample size, especially of control or reference organisms, to identify significant exposure. Apart from acetylcholinesterase, there are few chemical-specific biomarkers for evaluating pesticide effects in aquatic ecosystems.

Effects. If uncertainty in the hazard aspect of the risk assessment does not allow the conclusion that the reductions in exposure will result in reduced risk, then specific effects studies should be requested to improve the risk assessment. Such studies should focus on improving the prospective or predictive element of the risk assessment and are discussed in Chapter 2.

When considering the need for biological field studies to evaluate the efficacy of mitigation, the conceptual bases of predictive versus retrospective risk assessments need to be separated. Retrospective risk assessments assume there is an identified problem (*i.e.*, fish kills) that specific field studies can be designed to evaluate and address. However, if specific events are not known to occur, then designing retrospective field studies to "prove" that such effects do not occur is inappropriate. For example, if risk managers, based on input and guidance from scientists, decide that it is important to determine the secondary effects of a chemical or that due to the pesticide's properties other specific information outside the standard array of tests would significantly improve the risk evaluation of the chemical, then additional work may well be justified. In such cases, studies should be conducted in experimental systems to provide predictive information (as discussed in Chapter 2).

Although the exposure and toxicological components of predictive risk assessments are sound, field verifications of prospective risk assessments for pesticides have seldom been conducted. Regardless, field biological monitoring studies for validation of the risk assessment should never be substituted for a poorly defined risk assessment. Monitoring studies may be required if they are determined to be an effective (cost, efficiency) approach for gathering the required data. However, the need and use of the required data must be clearly articulated. Problems associated with high variability, lack of adequate reference sites, and difficulty in establishing cause-effect relationships must be defined before a specific biological monitoring requirement is considered. Field monitoring should never be considered routine and, if required, close coordination of risk managers and scientists in the problem formulation stage is essential. If there are specific uncertainties in the predictive risk assessment, laboratory or simulated field studies to improve the risk assessment should be considered before biomonitoring studies are required.

This point is illustrated in Figure 2-13 of Chapter 2: Risk Characterization, which indicates there is a specific probability of exceeding an atrazine exposure concentration which may adversely impact a specific percentage of the species exposed. Mitigation should attempt to reduce the probability of exposure such that exposure no longer overlaps with the distribution of effects. Due to uncertainty associated with the dose-response relationship, it is often unknown whether, in reality, the effects distribution is underprotective or overprotective. Given overlap of the exposure and effect curves and lack of information on field incidents, additional effects data should focus on improving confidence in the predictive capabilities of the toxicity distribution relationship. Biological monitoring studies will, under most circumstances, not result in an increased understanding of the potential for effects to occur. They may provide an assessment of the condition of the aquatic resource at risk; however, such conditions would be the result of numerous factors (habitat, nutrient levels, toxicants, erosional inputs, etc.).

As alluded to earlier, retrospective evaluation of specific incidents represents an example of when monitoring for adverse effects may be appropriate for the assessment of the efficacy of mitigation. In such instances, effects must be known to occur and there needs to be a stressed environment to be examined and measured (Suter, 1993). As an example, if recurrent fish kill incidents are a primary factor identifying major concerns for a specific chemical use, and mitigation is implemented to reduce exposure and eliminate such incidents, then it is logical to evaluate whether incidents are in fact eliminated. Specific examples of approaches that could be utilized include efforts to increase the efficiency of identifying fish kills when they occur or the use if in-situ bioassays to look at direct effects on fish. With the necessary modifications and after validation, other approaches (Karr's IBI) used in other regulatory arenas may also be appropriate.

Relative to incident reporting, registrants could include information on product labels that would both require and aid users in reporting of incidents that may be observed following pesticide use. By implementation of a broad-scope program of this nature, USEPA and registrants could cooperate to greatly upgrade current identification and reporting systems for aquatic exposure incidents. Currently, no comprehensive system is in place that provides for the identification and reporting of pesticide-related environmental incidents. Implementation of a program could significantly improve the knowledge base on the actual occurrence of incidents.

Although individual chemicals can contribute to population, community, or ecosystem-level changes in aquatic environments, a number of other stressors (habitat modification, soil erosion, nutrient additions, and other human activities) can also impact aquatic populations. Because of these factors, any decision to use field biomonitoring approaches must be carefully considered. The development of field procedures must integrate both effects and exposure components such that cause-effect relationships can be addressed.

Field biological monitoring has been utilized in state and federal water quality investigations, although it has not been applied within the FIFRA regulatory arena. Experience gained by other regulatory agencies in the use of biological monitoring should be considered when developing or evaluating the applicability of such studies to FIFRA regulatory needs. Studies must be designed that are known to provide scientists and risk managers with the information necessary to address specific regulatory questions or concerns. Diligence in requiring such studies is necessary to ensure all appropriate questions associated with problem formulation are adequately addressed. Specific factors that must be thoroughly considered before application of biomonitoring studies under FIFRA include

- the realization that approaches and methodologies which are still in the research phase or are used within other regulatory arenas cannot be directly applied within the FIFRA regulatory framework without proper validation.

- the realization of the basic difference between research projects designed to further our understanding of potential impacts and regulatory requirements necessary for informed risk management decisions.
- the need to clearly define the regulatory endpoint prior to implementing specific biomonitoring projects. The risk manager must clearly articulate how the information will be used to evaluate the efficacy of mitigation or improve the risk assessment.
- the ability of the study design to establish cause-effect relationships and the study's power to determine effects on specific ecosystem components identified as important by the risk managers.

Two basic approaches for field biological monitoring are *in-situ* toxicity tests and biological surveys.

***In-Situ* Laboratory Bioassays or Toxicity Tests.** The objective of such studies is to evaluate on-site effects, such as the frequency and extent of mortality, associated with real-world exposure conditions. Such studies must be designed to establish cause-effect relationships.

Approaches include *in-situ* toxicity tests using caged organisms and bioassays with site water removed and tested in the laboratory. The advantages and disadvantages of *in-situ* testing include the following.

Advantages

- Actual exposure and associated, quantifiable effects are included.
- Cause-effect relationships could be established; however, conditions necessary to establish these relationships are often beyond the control of the investigator.
- Methods are generally available, although they need to be modified and focused for site-specific conditions.

Disadvantages

- Random mortality occurs; associated false positives are difficult to refute.
- Maintaining healthy organisms in the cages is difficult.
- Cause-effect relationships are difficult to establish.
- Extensive analytical work is required to establish cause-effect, including the analysis of all potential stressors other than the pesticide of interest.
- If multiple sites are required, testing can be expensive.
- Reference or control sites are difficult to find.
- Representativeness of sites is important and difficult to address, especially for widely used pesticides.
- Testing is dependent on stochastic events (drift, rainfall) for exposure to occur, making such studies difficult and very expensive to conduct.
- Spatial and temporal variability can make interpretation of results difficult.

Biological Surveys (Indices, Biocriteria, Rapid Bioassessment Protocols [RBPs], Population and Community Studies). The objective is to evaluate the current "biological integrity" or "relative condition" of the resource associated with exposure to a specific pesticide or application site. A key criterion for the success of such an approach is a study design that can differentiate the effects of a single pesticide from the other stressors affecting the biota of the ecosystem. The ability of such studies to meet this criterion needs further validation for pesticides in aquatic habitats within ecosystems.

When Phase 3 studies have been initiated in response to incident reports or to biological monitoring indications of resource decline, a more focused study can be designed. Such a study might include a site-specific comparative analysis of recruitment, mortality, and condition in a particular fishery; evaluation of key life history elements (for example, contamination of spawning grounds); or studies of food chain dynamics or contamination.

Advantages

- Effects are associated with actual exposure.
- The studies fit within the current USEPA framework/policy of holistic, watershed approaches.
- Over time, benchmarks on ecosystem health will be established.
- These studies can possibly fit within and supplement ongoing state and federal programs.
- The status of the resource is directly evaluated.

Disadvantages

- Natural variability must be known/established to appropriately design a study and interpret the results.
- Predictability, and thus extrapolation to other areas, is poor.
- It is very difficult to separate the impact of a single stressor and establish a cause-effect relationship.
- There are extensive analytical requirements for identifying cause-effect relationships.
- Methods that are directly or immediately applicable to single chemicals are not available/validated.
- Appropriate reference sites are lacking.
- Representativeness covering all use areas is lacking.
- The stochastic nature of events controlling exposure can make the conduct of such studies difficult.
- Establishing statistical confidence in results can be difficult given variability, etc.
- No baseline information is available for evaluating the effectiveness of mitigation.

- The scientific return on investment is low; cost is high given the data obtained.

Example of Phase 3 Mitigation Verification. As in the Phase 2 Section 5.2.2, the following example of the rationale associated with implementing Phase 3 requirements may be useful.

Based on the risk characterization, pesticide used on commercial turf and golf courses has exceeded an LOC for acute and chronic effects on both fish and aquatic invertebrates. The risk characterization predicts the potential for fish kills, and these predictions are confirmed by the incident data. The incident data have been reported over the last 10 years and are associated only with some of the geographical regions. The product is generally applied 2 to 4 times per year on a 14-day interval and is restricted to ground application only. The pesticide is fairly water soluble and has a low partition coefficient, so transport to the aquatic environment occurs primarily by runoff in the soluble phase. Erosional losses from the turf sites are not a concern.

Mitigation to reduce risk below the LOC, and thus eliminate or minimize fish kills, included

- a 75-foot buffer zone for application adjacent to bodies of water,
- a requirement to implement use and irrigation practices that control or minimize runoff volume, and
- a restriction on the use of products when conditions are conducive to runoff and rainfall is imminent. It may be necessary for such restrictions to be state-specific and may require prescription use.

In addition to the uncertainty associated with the ability of the mitigation measures to actually reduce exposure, there is also uncertainty associated with the extent of the fish kills that are occurring and whether they are occurring for all use patterns. Given this, verification may require components of both Phases 2 and 3.

Phase 2 Small-scale runoff studies providing predictive information on the reductions in runoff associated with the mitigation measures

Phase 3 Monitoring addressing (1) the extent of the fish kills associated with the specific use patterns that have a history of fish kills and (2) whether the exposure is great enough for similar fish kills to be associated with the use patterns without a history of incidents. Specific use patterns could be eliminated as a concern based on consideration of "landscape" exposure factors such that actual watershed monitoring may not be necessary.

A primary uncertainty is determining why fish kills are associated with specific use patterns and not other use patterns. Once these factors are understood, implementing additional mitigation options to address these factors may be possible.

5.2.4 Evaluating Long-Term Impacts of Pesticides

The development of procedures for evaluating long-term impacts of pesticides was a specific need identified by USEPA's Assistant Administrator's Ecological Effects and Environmental Fate Task Force. A variety of biomonitoring tools exists for such evaluations (for example, RBPs, biological indices); these tools have not, however, been convincingly validated or specifically adapted for evaluating individual pesticides in aquatic habitats. Validation and development work is required. When conducted, studies should be designed around the structure and function of ecosystems characteristic of the use site. It is important to develop field data establishing the relative condition of aquatic ecosystems in application areas. When it is possible to link field effects to individual pesticides, such assessments should be done. However, linking long-term evaluations of ecosystem integrity with pesticide-specific assessments or with evaluating mitigation effectiveness for specific pesticides would be very difficult. Concerns relative to long-term impacts should be integrated into existing or planned programs within USEPA or other federal and state agencies. This practice will help ensure data are generated in a coordinated, focused fashion that addresses the primary areas of concern associated with use patterns.

Existing natural resource monitoring programs (for example, programs of the Geological Survey, the proposed National Biological Survey of the Department of the Interior, and USEPA's EMAP program) must be relied upon for identification of potential pesticide damage problems that may arise even under ideal conditions of safety evaluation and mitigation controls. In such circumstances, investigation of the causes of resource impairment requires a knowledge of the source strength, persistence, mobility, and ecotoxicology of possible pesticides. The USEPA program of modeling and monitoring conceived under the new paradigm will significantly accelerate the process of identifying and removing those pesticide stressors—whether single or synergistic—that may escape the regulatory net.

5.3 VERIFICATION RECOMMENDATIONS

1. To ensure consistent and useful mitigation and monitoring efforts, improved collaboration and communication between the industry, USDA, USEPA, environmentalist user, applicator, and other federal agencies needs to be developed and formalized.
2. As part of the collaborative activities between agencies and industry, existing data and understanding on mitigation practices (or BMPs—best management practices) need to be organized into data or "knowledge" bases that can be easily utilized and easily updated as new information is developed.
3. Where deficiencies in information are found to exist for both the effectiveness of "on-site" mitigation practices as well as understanding the "field-to-stream" transport of pesticides, including processes and parameters necessary for mathematical modeling, a list of research needs requiring funding should be identified.
4. A comprehensive program for the identification and reporting of pesticide-related environmental incidents by the users should be required.

REFERENCES

Suter, G.W., II. 1993. *Ecological Risk Assessment*. Lewis Publishers, Chelsea, MI.

Chapter 6: LONG-TERM CONSIDERATIONS

As currently practiced, the ecological risk assessment process used in pesticide regulation represents a pragmatic approach to direct toxic risk to nontarget organisms. Acute and chronic toxicity measurement endpoints are compared to environmental fate and exposure measurement endpoints to develop risk quotients in a prospective risk assessment approach. Until recently, field studies and simulated field studies (**mesocosms**) were sometimes used to verify or rebut a presumption of risk predicted by the risk quotient method. These field studies employed exposure scenarios typical of use patterns under consideration for registration, and were generally conducted for a single growing season. Following registration, additional determination of ecological problems caused by pesticide use is largely restricted to incident reports of fish and wildlife poisonings.

An implicit (although perhaps rarely stated) assumption in the above risk assessment approach is that laboratory studies and post-registration incident reports are sufficient to estimate and verify major ecological risk concerns for pesticides. The current approach may be adequate for preventing "widespread and repeated mortality" as a primary risk management goal. However, equating the elimination of highly visible, fish-kill incidents with long-term protection of aquatic ecosystems is ecologically naive. Long-term changes in aquatic ecosystem structure and function may certainly occur with repeated and widespread acute toxic insults. Yet, less obvious indirect effects over time may have equally dramatic consequences over the longer term. Interestingly, the recent mesocosm study approach provided some information on alterations to aquatic community structure and function, but the results were restricted to a single growing season. The effect of a single pesticide applied annually over multiple years, or effects of all pesticides from different crops and different application windows in heavily used watersheds, remains little studied.

Reconciling the current ecological risk assessment process for pesticide registration with longer term ecological risk analysis and monitoring approaches is difficult. Obviously, any risk assessment situation has practical limitations in terms of time and resources. Not surprisingly, fate- and toxicity-based risk assessment approaches are more highly developed than true long-term ecological assessment approaches. Fate- and toxicity-based risk approaches also represent a more tractable scientific analysis problem than long-term ecosystem assessment approaches at the current time. The reality of this situation is reflected in the deliberations and recommendations of the Aquatic Risk Assessment and Mitigation Dialogue Group (Aquatic Dialogue Group).

Tangible discussions and recommended refinements for pesticide risk assessments dealing with environmental fate and effects methods selection and data analysis occurred relatively quickly in the Aquatic Dialogue Group working sessions. The ecotoxicological emphasis on pesticide risk assessment is historically based and provides cost-effective risk estimates of localized exposure and short-term temporal effects of pesticide to aquatic resources. Nevertheless, the laboratory-based approach is not entirely capable of resolving all ecological risk concerns that may arise from pesticide use over the longer term.

In an ideal scientific and regulatory scenario, ecological risk assessment processes for pesticide registration would include a number of salient ecological assessment and measurement endpoints, as well as mitigation actions necessary to eliminate unwarranted negative ecological consequences resulting from pesticide use. However, deciding what ecological assessment endpoints are "important," what level of adverse change is "unacceptable," how to adequately measure assessment endpoints, linking adverse effects to a pesticide or combination of pesticides, and delineating pesticide effects from other environmental stressors typical of dynamic ecological environments are all challenging scientific problems. In addition, how to develop and implement long-term effects and ecological monitoring data under existing Federal Insecticide, Fungicide and Rodenticide Act (FIFRA) mandates and regulatory policies is not currently well defined. Nevertheless, strategies for determining the long-term ecological impacts of pesticides were a specific need identified by the Office of Pollution Prevention and Toxic Substances (OPPTS) Ecological Effects and Environmental Fate Task Force (Appendix A).

Considering the scientific, regulatory, legal, and public opinion processes that enter into pesticide regulatory actions, it is not surprising that Aquatic Dialogue Group members had divergent opinions about the scope, scale, and application of long-term ecological risk and monitoring strategies under FIFRA. Toxicologically based risk assessments are certainly valid within their methodological assumptions. However, such risk assessments do not encompass long-term risk to broader structural and functional aspects of aquatic ecosystem integrity. Although there is a scientific and regulatory basis for toxicology-based risk assessment in pesticide risk assessment, the assessment and measurement endpoints currently used do not always reflect current thinking and emerging ecological risk assessment trends developed in other federal and state regulatory program areas and natural resource management. Examples include ecosystem monitoring and assessment approaches focused on populations and communities, a move toward larger landscape areas such as watersheds, and cumulative impacts of multiple stressors such as total pesticide loadings for all sources in a watershed or region. Not surprisingly, long-term ecological risk characterization requires substantially different scientific strategies, approaches, and methodologies than are now currently embodied in short-term toxicological risk characterization schemes.

The toxicology-based approach used for ecological risk assessment is fairly well defined, understood, and routinely applied in the pesticide regulatory arena. However, results of toxicologically based risk assessments cannot be extrapolated beyond their inherent methodological assumptions and limitations. Certain aspects of pesticide risk to ecosystems are clearly beyond the scope of routine toxicological risk assessments. Similarly, emerging long-term ecological risk assessment approaches hold promise for the future, but often require additional research and verification prior to incorporation into legislation and regulatory policy. In any event, it appears likely that prospective toxicological-ecological risk assessment will remain prominent in pesticide regulatory operations for the foreseeable future, with broader based ecological assessment and monitoring approaches entering the picture as methods and processes are developed and verified.

The Aquatic Dialogue Group recognized certain ecological uncertainties attendant with basing risk assessments solely on laboratory-based fate and effect studies, but also acknowledged their historical significance and utility in pesticide regulatory decision making. Broader incorporation of ecosystem-based risk assessment approaches was generally viewed as a positive development, but requires careful integration of research, risk assessment, and risk management. Some Aquatic Dialogue Group members pointed to the development of ecological risk assessment and monitoring guidance developed for other U.S. Environmental Protection Agency (USEPA) regulatory offices as being potentially useful for pesticide monitoring (*e.g.*, ecological risk assessment guidance for hazardous waste sites, Office of Water biocriteria, sediment criteria, wildlife criteria, and Rapid Bioassessment Protocols [RBPs]). Other participants, however, believed they have not as yet been convincingly validated or specifically adapted for evaluating individual or groups of pesticides in aquatic habitats. Validation and development work is required, and studies on the structure and function of ecosystems characteristic of use patterns may be more directly applicable to pesticide problems.

One common theme in the discussions was the need to develop a broader perspective on ecosystem issues within the pesticide regulatory arena. Ecosystem protection requires a fundamental shift in thinking from the protection of individual component resources (fish, aquatic invertebrates, etc.) to protection of the underlying life support systems that nurture the various component resources. Similarly, promoting an attitude of proactive management and protection of natural resources may prove to be more cost efficient than taking management action only after the appearance of recurrent problems (for example, repeated fish kills) associated with a particular pesticide use. A key aspect of developing a broader ecological perspective lies in the area of better communication and understanding between rapidly evolving research developments, definition of research needs for pesticide risk assessment applications, and risk management questions in pesticide regulatory decision making.

One area identified by the Aquatic Dialogue Group as needing greater attention is the dynamic interaction of basic research, applied research, and the application of science in a legislatively mandated ecological risk assessment arena such as pesticides. As discussed in the Introduction of this report, specific roles for risk assessors and risk managers are emerging in pesticide regulatory decision making. However, superimposing the role of research and the inherent lag time between the development of scientific advancements and their routine application in risk assessment and risk management scenarios is rarely delineated.

Although some regulatory procedures are science based (*e.g.*, risk assessment), others may be primarily economic, social, and political (*e.g.*, risk management). As a result, advancements in basic science, applied science, regulatory needs, and regulatory policy often occur independently with little consideration of their interdependence. A model for better integration of risk assessment and risk management has been promoted by the USEPA Risk Forum (see Figure 2-3, Chapter 2), but does not really address the role of research in the overall process. One possibility for superimposing research on the integrated risk assessment/risk management is shown in Figure 6-1. Better communication between researchers, risk assessors, and risk managers would allow for better understanding of all the salient issues that ultimately go into regulatory decisions.

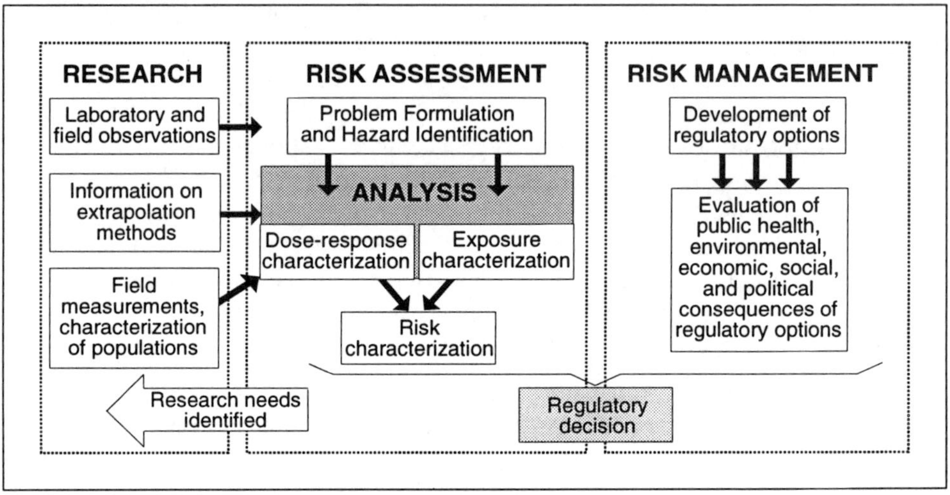

Figure 6-1. The risk assessment process as used by a number of regulatory authorities

An interactive research, risk assessment, risk management process would assist the development of regulatory science, as well as the overall quality and consistency of regulatory decision making. An approach of this nature goes significantly beyond mere fine-tuning of predetermined regulatory risk assessment schemes, and may begin a fundamental transformation of processes required to ensure that sustainable agriculture and sustainable aquatic resources are not mutually exclusive goals. In any event, environmental protection is ultimately dependent on a variety of research, risk assessment, and risk management activities. An open and inclusive process that considers various points of view will ultimately provide better information concomitant with increased environmental protection. It should be emphasized that such participation must include users as well as environmental advocacy groups to be successful.

CONCLUSIONS FOR ECOSYSTEMS ISSUES

1. Risk manager/risk assessor communication and understanding requires further development in the ecological area and should balance the concepts of sustainable agriculture with sustainable natural resources.
2. Focus on pesticide risk assessment has been based on toxicological analysis, which is currently a more tractable analysis problem than broader ecosystem concerns.

3. The current prospective risk assessment process is adequate to estimate the risk of direct acute and chronic risk to major groups of aquatic organisms, but does not account for indirect effects that may affect the structural and functional integrity of aquatic ecosystems.
4. The primary problems caused by pesticides result from off-site movement of pesticides where they affect nontarget resources. Pesticide-induced problems may be localized to user sites or affect larger spatial scales from regional use. Assessment and measurement endpoints for these extremes require substantially different data and methods evaluation.
5. Prospective ecological risk assessments are defined by the inherent methodological assumptions of scope, scale, and assessment and measurement endpoints implicit in their design and should not be extrapolated beyond their scientific assumptions.
6. Ecosystem protection requires a shift in thinking from protection of individual component resources to protection of the underlying processes that ultimately sustain both managed and natural ecosystems. This shift also requires a move away from resolution of existing problems to preventive maintenance.
7. Current ecological risk assessment schemes for pesticides would benefit from a better understanding of the following:
   - Land-use pattern with respect to the site-specific environments must be considered.
   - The incremental effect of pesticides on other ecosystem stressors requires definition.
   - Overall loadings of pesticides in ecosystems from all sources are rarely considered and poorly understood.
8. Ecosystem approaches require large-scale data collection and analysis and are best conducted by cooperative agreements between investigators, institutions, and agencies. Linkages between applied and natural-resource scientists, academic institutions, and federal programs and agencies should be developed to maximize resources and information on a collaborative basis.
9. Management of aquatic ecosystems is currently conducted in venues other than FIFRA regulatory actions, but pesticide regulatory decisions and mitigation actions should be compatible with other environmental protection statutes.

UNITED STATES ENVIRONMENTAL PROTECTION AGENCY
WASHINGTON, D.C. 20460

OCT 29 1992

OFFICE OF
PESTICIDES AND TOXIC
SUBSTANCES

MEMORANDUM

SUBJECT: Decisions on the Ecological, Fate, and Effects Task Force

FROM: Linda J. Fisher
Assistant Administrator

TO: Douglas Campt, Director
Office of Pesticide Programs

In May 1992, I initiated the Ecological, Fate and Effects Task Force to evaluate the testing requirements for ecological and environmental fate with respect to their impact upon OPP's regulatory programs. For the past several months, numerous OPP managers and analysts, as well as participants from OPPE, OPPT, OGC, and ORD have put an extraordinary amount of time and effort into this workgroup. These decisions, once implemented, will result in protective and timely decisions in the area of ecological risk management. The attached document outlines the specific decisions on the very important issues of ecological risk management raised by the Task Force. There are still many details to resolve as this new paradigm is implemented. I have asked Deputy Assistant Administrator Vic Kimm to chair an implementation committee.

The effort put into the Task Force deliberations was clearly monumental. The Program showed a remarkable commitment to analyzing and addressing these issues. In addition, the support provided by OPPT, OPPE, ORD, and OGC was outstanding, as were their contributions to our policy discussions.

The principle issue addressed by the Task Force is the effect of the data generated from OPP's ecological and environmental fate data requirements on the Reregistration program. As a regulatory agency, we must have a clear sense of how to use the data which we require. The value added by these data on regulatory decisions (registration and reregistration) was carefully reviewed. Secondly, the Task Force evaluated the effect of collecting and reviewing ecological and environmental fate data on the legislative deadlines established for reregistration. Although the Task Force focused much of its attention on reregistration, many of its findings apply to the registration program.

In our registration and reregistration programs, the Agency is interested in setting risk based priorities and making protective decisions in a timely fashion. We must recognize that registration is a continuous process. Registration decisions and reregistration decisions made under FIFRA 88 may be revisited in following years as we learn more about ecological risk.

In summary, regulatory managers will attempt to make decisions regarding ecological risk earlier in the data gathering process, often with less data than are currently required. Decisions should be made in the absence of the higher tiered data requirements whenever possible. In addition to making decisions earlier in the data gathering process, OPP will rely much more extensively on risk mitigation for ecological effects when information indicates that some "level of concern" has been exceeded.

More specifically, OPP will no longer require avian and aquatic field testing, except in unusual circumstances. Rather, decisions will be based on lab testing, incident data and other information which can easily be collected to enable the program to better characterize potential risk. Ground water monitoring studies will be required only if risk managers believe that the existing groundwater database, including modeling analyses, do not provide an adequate basis for up-front regulatory decisions. Finally, the Agency will begin to develop a longer-term strategy for obtaining information needed to reduce uncertainty in evaluating ecological risk associated with long-term effects of pesticide use.

The decisions outlined here will tremendously improve our ability to meet the deadlines established for reregistration in FIFRA 88, and to evaluate registration actions in a more timely fashion. More importantly, I believe these changes will result in a significant and prompt improvement in environmental quality. More decisions -- environmentally protective decisions, with a major focus on up-front remediation where appropriate -- will be made with far greater speed under the new paradigm than would have ever been feasible under the existing testing scheme.

Our work has only begun. We now must develop a detailed implementation plan to carry out this new paradigm. This plan should include specific process changes with respect to both registration and reregistration, disposition of existing data call-ins (DCIs) which are affected by this effort, development of a process for risk mitigation, and specific internal procedures to ensure the exposure estimates and the estimated environmental concentrations (EECs) are calculated using appropriate modelling and actual field data when feasible. Clearly, there are numerous issues which will need to be addressed, beyond those identified here, in the implementation plan. It is my goal to have a final implementation plan by January 5, 1993.

Again, I wish to thank you and your staff for the commitment you have brought to this very important effort.

ATTACHMENT: *Program Guidance on Ecological Risk Management*

I. AQUATIC (mesocosms)

1. Assistant Administrator (AA) Decision

The AA supports the workgroup recommendation to make regulatory decisions on aquatic risk in the absence of the mesocosm studies. Reregistration, Registration, and Special Review decisions over the course of Reregistration will be based upon information available from laboratory studies, published information, and incident data. Regulatory decisions will be based on the effects, both acute and chronic, of pesticide use on several aquatic species (warm water fish, cold water fish, marine fish, invertebrates, estuarine/marine mollusks, and estuarine/marine shrimp). Although the Agency believes that long-term, indirect effects of pesticide use on aquatic ecosystems may be important, the Agency does not have a testing scheme in place to accurately measure such effects within the time specified for Reregistration. For policy reasons, the Agency has defined a set of regulatory endpoints for reregistration purposes. As we develop a better understanding of these endpoints, such as the indirect effects of pesticide use on aquatic ecosystems, the Agency reserves the right to regulate against such endpoints in the registration program in the future.

OPP will begin to develop a strategy for making regulatory decisions on the long-term ecological effects of pesticide use. In this effort, OPP should consider a longer time frame for field studies as well as the possibility of testing a single representative of a class of pesticides. The current field studies have a duration of about two years and are performed on a chemical by chemical basis. These studies do not provide risk managers with information which greatly enhances the risk management decision-making process.

For future reregistration and registration purposes, OPP should use risk mitigation to the extent feasible to achieve acceptable risk. Risk mitigation includes all practices which reduce hazard or exposure. Mitigation can be in the form of reduced application rates, mandatory tillage practices or a change in the pesticide's formulation, to name a few. With respect to the current aquatic triggers of "levels of concern", the AA supports the workgroup's recommendation to continue using the existing triggers. However, risk assessments should include a small stream as the receiving body of water. Risk management decisions should be based on the effects to flowing bodies of water unless a risk case, on a use specific basis, can be made as to the relevance of protecting small, closed water systems (such as a pond, wetland, or a small lake).

2. Levels of Concern

ACUTE:

If the EEC > 1/2 LC50 (LC50 as measured for the most sensitive species) (EEC is estimated environmental concentration; LC50 is the lethal concentration which kills 50% of the test species), the aquatic risk of a pesticide is deemed to be of high concern, and may warrant regulatory action in addition to restricted use classification.

If 1/10 LC50 <= EEC <= 1/2 LC50, then the pesticide is considered for classification as a restricted use pesticide.

If EEC < 1/10 LC50, then the pesticide has low aquatic risk and no additional regulatory action will be pursued.

CHRONIC:

For both the Chronic Aquatic Invertebrate and Fish tests and the Life Cycle Fish tests, the level of concern is reached if the EEC >= LEL (LEL is the lowest effect level). These are both reproductive tests.

QUALITATIVE FACTORS:

Incident reports of fish kills trigger a level of concern. Availability of alternative should be reviewed.

3. Effect on Reregistration

How will this policy change affect decision-making in the reregistration program? OPP will make regulatory decisions for aquatic risk following the guidelines established in the Levels of Concern (LOC) project. If a risk exceeds the LOC, the risk assessment should be refined using standard approaches such as application of more sophisticated models to estimate exposure and the EEC. If the LOC exceedance is supported, risk managers should require or negotiate risk reduction measures. It may be necessary to calculate the cost of mitigation in a very crude form prior to negotiating risk mitigation. This will help the Agency decision makers to make better informed decisions regarding mitigation measures to pursue in discussions with registrants. If the LOC is still exceeded after risk reduction measures are evaluated, a risk/benefit based decision, leading to specific regulatory determination, is necessary. A risk/benefit determination is necessary at this point because the concept of "acceptable risk" takes into consideration benefits. Thus acceptable risk may exceed the level of concern, and cannot be defined with a simple numerical standard. To define acceptable risk a preliminary benefits assessment must be conducted. If it becomes clear that an in depth review of the risks and benefits is necessary to make a determination, then the chemical will be declared ineligible for reregistration, pending a Special Review or other regulatory action.

For purposes of risk characterization, OPP will evaluate the following parameters which are viewed by the AA as essential pieces of information in making regulatory decisions in the area of aquatic risk. 1) The extent of use; 2) The environmental fate of the chemical; 3) The quality of the water body and surrounding use areas as habitat; 4) The species exposed; 5) incident reports of fish (or other aquatic species) kills; and, 6) availability of substitutes, including biological controls and integrated pest management. Each of these provides important qualitative or quantitative information to the risk manager.

Evaluating the risk reduction measures will be an important new step in this process. The final RED (Reregistration Eligibility Document) decision will rely significantly on EFED's (Environmental Fate and Effects Division) ability to evaluate proposed risk mitigation in a timely fashion. SRRD (Special Review and Reregistration Division) will also be required to make risk/benefit based reregistration decisions, often with uncertainties. OPP may need to consider the feasibility of follow-up residue monitoring to ensure that risk mitigation is successful. SRRD must continuously evaluate the contribution of additional information with respect to their ability to make decisions. New information should only be required when it significantly improves our ability to evaluate the risks and benefits of a pesticide.

4. Status of Mesocosm DCIs (Data Call-Ins)

At this point, complete, acceptable mesocosms will not normally be required for purposes of regulatory decision making (reregistration or registration), as these studies do not generally provide regulatory managers with information to make better regulatory decisions. RED decisions can and should be made in the absence of mesocosm studies. If SRRD or RD (Registration Division) feel that a regulatory decision cannot be made in the absence of an aquatic field study, the mesocosm can be required. OPP will continue to accept mesocosm studies which registrants submit to the Agency. These studies will be evaluated and used as confirmatory information in the aquatic risk assessment to the extent that the mesocosm provides relevant information with respect to the aquatic levels of concern. If the registrant believes that a mesocosm will refute a presumption of risk, they can pursue such a study. However, the Agency does not need to wait for such a study before making regulatory decisions.

5. Future Directions in Aquatic Risk Assessment/Risk Management

Over the next five to ten years (period of FIFRA 88 reregistration) OPP should aim to greatly reduce the direct acute and chronic risks to aquatic environments. During that timeframe, OPP should develop a regulatory scheme for protecting aquatic ecosystems from the long-term effects of pesticide use based upon improvements in our evolving scientific understanding of ecosystems. This may include the development of a long-term, mesocosm-like study, or actual residue monitoring of water bodies. OPP should evaluate all options which provide the Agency with the type of information which will allow us to make future regulatory decisions in a cost-effective manner.

II. AVIAN

1. AA Decision

The AA agrees with the workgroup conclusion that the avian field study provides very limited new information to an avian risk assessment, and that such field studies confirm the results of the lab studies. Except in unusual circumstances, where the value of an avian field study to a regulatory decision maker is significant, the avian field studies will no longer be required for purposes of reregistration and registration of individual active ingredients.

2. Levels of Concern

ACUTE:

If the EEC > 1/2 LC50, or the LD50/sqft > 1/2, the avian risk of a pesticide is deemed to be of high concern, and may warrant regulatory action in addition to restricted use classification.

If 1/5 LC50 <= EEC <= 1/2 LC50, then the pesticide is considered for classification as a restricted use pesticide.

If EEC < 1/5 LC50, then the pesticide has low avian risk and no additional regulatory action will be pursued.

CHRONIC:

For chronic avian tests, the level of concern is reached if the EEC >- LEL.

QUALITATIVE FACTORS:

Incident reports play a critical role in the avian risk management process. Also important are extent of use, proximity to high value habitat, the fate of the chemical, the species exposed, and the availability of safer substitutes.

3. Effect on Reregistration

RED decisions will be made in the absence of avian field studies. If the chemical exceeds a level of concern for an avian endpoint, OPP will follow the same basic framework as for aquatic risk management. First, OPP will refine its hazard and exposure calculations. Secondly, OPP will evaluate risk reduction techniques (including the costs of mitigation). If mitigation is insufficient to reduce the risk to below the level of concern, a preliminary benefits assessment will be preformed and a risk/benefit determination will be made. Although the level of concern identifies a point at which the Agency is unable to dismiss potential risks, it does not define either acceptable risk or the standard by which final risk-benefit determinations will be judged.

As stated in previous avian decisions, the Agency will not accept widespread and repeated avian mortality in the face of minor economic benefits to society. The "widespread and repeated" standard can be met on a local or regional as well as a national basis. As the economic benefits of a chemical increase, the standard for significant economic benefits may not be outweighed by risks unless those risks are very high, very widespread, or involve especially valued species or habitat.

For purposes of risk characterization, OPP will evaluate the following parameters, which are viewed by the AA as essential pieces of information in making regulatory decisions in the area of avian risk. 1) The extent of use (acreage); 2) The environmental fate of the chemical 3) The quality of the crop and surrounding use/areas as habitat (wildlife utilization); 4) The species exposed; 5) The incident reports of bird kills; and, 6) The availability of substitutes, including biological controls and integrated pest management. Each of these provides important qualitative or quantitative information to the risk manager.

4. Status of Avian Field Studies DCIs

Avian field studies will not be required for a substantially complete data base except in highly unusual circumstances (the regulatory management divisions are unable to make a regulatory decision in the absence of such a study). Thus registrants will not be required to complete such studies.

5. Future Directions in Avian Risk Assessment/Risk Management

OPP needs to develop a long-term strategy for improving the scientific risk assessment methodologies, building on findings of the most recent avian risk dialogue group. In addition, OPP should begin to evaluate the viability and usefulness of longer term avian monitoring in high use areas as well as a research plan for better understanding the impacts of pesticide use on birds.

Avian and aquatic levels of concern (acute and chronic) have been established in regulation and guidance over the last 10 to 15 years. As far as the Agency knows, theses levels are appropriate, and OPP should continue to use these "triggers" for regulatory purposes. As a normal part of the scientific and regulatory process, these numbers will be subject to periodic review.

III. GROUND WATER

1. AA Decision

The AA supports making an explicit risk management decision before requiring a small scale prospective ground water monitoring study. The current practice is to require a small scale study if the chemical proves to be mobile and persistent, as determined by lab studies and a field dissipation study. There is a high level of confidence in the prospective ground water monitoring study from a scientific as well as a regulatory perspective. These studies can provide valuable information to regulatory managers in the protection of ground water from problem pesticides. As with all studies which the Agency requires, we should have a clear sense of how we will use the information, in a regulatory context, which the study will provide.

2. Levels of Concern

Mobility is defined as Kd <= 5 MG/gm or Koc <= 500 Mg/gm, or detection of the compound 90 cm in the soil profile in the soil dissipation study, and;

Persistence is defined as a soil half-life greater than 2–3 weeks, and;

A risk based (human health or ecological) trigger is exceeded, or has a high potential to be exceeded (OPP is currently working to define this trigger based on some fraction of a Maximum Contaminant Level (MCL) or Health Advisory (HA), and as appropriate, frequency of detections in high use areas).

3. Effect on Reregistration

The new process will involve a risk management decision before requiring the monitoring study. The following steps will be followed in the risk management process:

1) Determine if there is a known human health or environmental concern with the compound. If not, do not require the monitoring study; if so, go to 2);
2) evaluate the likelihood of the compound reaching a level of concern in a vulnerable area, given the entire data base of the compound (lab data, other monitoring, modelling, etc....); if it appears unlikely that the compound will

reach ground water at a level of concern, do not require the monitoring study; if it appears feasible or likely, go to 3);

3) if the existing evidence indicates that a pesticide exceeds the Agency's threshold (level of concern), then regulate without the monitoring study; if the evidence is equivocal, then go to 4);
4) if possible, identify mitigation measures which will reduce or eliminate the Agency's concerns, and then negotiate or require such measures as a condition for registration/reregistration; consider requiring monitoring in high use areas to evaluate mitigation measures over time;
5) if mitigation is not feasible or the registrant and the Agency cannot agree on feasible measures, either require the monitoring study or regulate based on existing information.

A separate workgroup within OPP is identifying the levels at which various regulatory actions are to be taken. The ground water workgroup will develop a logical, risk-based hierarchy of regulatory actions based upon levels of concern. Issues the workgroup will address include the following: What is the level of concern for a label advisory? Restricted Use classification? Risk mitigation or other regulatory restrictions? Under what conditions will we propose a State Management Plan?

The AA understands that for registration purposes, there will be little or no existing monitoring data to guide OPP in making a determination about the potential of a chemical reaching groundwater. Thus the need for a prospective monitoring study for registration purposes may be greater than it will be for reregistration purposes, due to the greater level of uncertainty.

4. Status of Existing Monitoring DCIs

OPP should perform a cursory review of the 45 chemicals which already have ground water monitoring DCIs to determine if a regulatory decision could be made in the absence of such studies. In particular, if the existing data base on a chemical indicates that regulatory action is warranted, we should pursue such action, and classify the monitoring study as optional. Where such studies are submitted to OPP, they should be evaluated and considered confirmatory. For chemicals which do not have a ground water DCI, we should go through the five-step process outlined above.

IV. RISK MITIGATION

1. AA Decision

The Assistant Administrator fully supports the concept of using risk mitigation for pesticides which pose risks of concern. Over the next few weeks OPP should develop a process which details how and when mitigation will be used. Of greatest importance should be how the program will evaluate risk reduction measures. This plan should also identify where in the risk management process mitigation will be imposed as well as who should be responsible for identifying appropriate measures. For both avian and aquatic risk mitigation, OPP should consider post-mitigation monitoring to evaluate the effectiveness of the mitigation measures.

Risk mitigation should be considered whenever a level of concern is exceeded to reduce exposure or hazard and provide added protection for the environment at the earliest possible time.

V. EXPOSURE, FATE, AND ESTIMATED ENVIRONMENTAL CONCENTRATION

1. AA Decisions

The Assistant Administrator strongly supports efforts to move toward more realistic exposure estimates based on the most realistic and sophisticated EEC calculations using our most qualified personnel and employing the best available data and modeling techniques. In addition, OPP should use actual residue data in its ecological risk assessments whenever possible. Kenaga data should be used only when appropriate actual residue information are not available. OPP should evaluate the possibility of enhancing it existing residue chemistry data requirements to include the type of residue data used in ecological risk assessments.

VI. RISK MANAGER - RISK ASSESSOR INTERACTION

1. AA Decisions

The Assistant Administrator concurs with the workgroup's findings that there is a lack of guiding principles, objectives, and policies to guide risk assessment and risk management in the ecological area. This document should serve as a first step in providing such guidance. In addition, the Assistant Administrator finds that the roles of the risk assessors and risk managers in the area of ecological effects are not as clearly defined or well understood as they should be. OPP should begin to work through the recommended options with respect to risk assessment/risk management interaction which are identified in the risk assessment/risk management chapter.

VII. IMPLEMENTATION

OPP should develop a workplan which describes the process by which these decisions will be implemented. An implementation plan should be in final form by January 5, 1993. An implementation committee should operate under the direction of the Deputy Assistant Administrator of OPPTS. A preliminary plan and schedule should be presented to the AA by December, 1992. In the interim, program managers should strongly consider the direction of this guidance when making regulatory decisions in the areas of ecological risk management and assessment.

UNITED STATES ENVIRONMENTAL PROTECTION AGENCY
WASHINGTON, D.C. 20460

AUG 25 1993

OFFICE OF
PREVENTION, PESTICIDES
AND TOXIC SUBSTANCES

MEMORANDUM

SUBJECT: Implementation Paper for the New Paradigm

FROM: Douglas D. Campt
Office Director
Office of Pesticide Programs (H7501C)

TO: Victor J. Kimm
Acting Assistant Administrator

As you know, the policy decisions of the Ecological Effects and Environmental Fate Task Force were published in October 1992. The Office of Pesticide Programs formed a Workgroup in response to these policy decisions, which was charged with developing a plan to implement these decisions.

As agreed upon our briefing of July 28, we have completed our paper that describes the implementation plan and are forwarding it to you. The plan has been written to provide people both inside and outside the Agency with a conceptual outline of the process. Detailed implementation steps have purposefully not been included to allow the process to remain dynamic and evolve over the course of the next year.

We have had a great deal of interest in the implementation plan and are eager to distribute this document to people both within and outside the Agency. As soon as we receive your approval, we will begin this process.

If you have any questions, feel free to call me.

Attachment

Implementation of the Policy Decisions of the Ecological Effects and Environmental Fate Task Force

I. Introduction

In March 1992, Linda Fisher, then Assistant Administrator for the Office of Prevention, Pesticides, and Toxic Substances of the Environmental Protection Agency (EPA), established an Ecological, Fate, and Effects Task Force. The objective of this Task Force was to review and assess the role of ecological and environmental fate data and data requirements on registration[1] and reregistration decision-making.

Specifically, the Task Force focused on how required ecological and environmental fate data are used in the risk assessment and risk management processes and how these data add to registration and reregistration decisions. In addition, the Task Force evaluated the impact of these data requirements on the timeliness of congressionally mandated reregistration deadlines as well as on the registration process.

In October 1992, the policy decisions of the Task Force were published (Ref. 4), and the Implementation Workgroup was formed to develop a plan to implement the "new paradigm". This document describes the implementation plan developed by the Workgroup, which is intended to provide people both inside and outside the Agency with a conceptual outline of the new process. Detailed implementation steps have purposefully not been included to allow the process to remain dynamic and evolve over the course of the next year. More detailed guidance will become available as the process evolves.

This document is organized into six sections. Section II provides background information and summarizes the policy decisions made by the Task Force. Sections III and IV describe the conceptual plan to implement the new paradigm for ecological effects and ground water, respectively. Section V provides clarification and responses to questions that have arisen regarding the new paradigm. The document ends with Section VI, a conclusion, which identifies long-term research goals, and Section VII which provides references.

II. Background

A. Definitions

Three components of the decision-making process under the new paradigm are risk assessment, risk management, and risk mitigation. These components are defined as follows:

---

[1] For the purposes of this paper, registration refers to the registration of new pesticides as well as the addition of new uses to already registered chemicals.

- Risk assessment is the scientific phase of the overall process and consists of hazard identification, and exposure assessment, ultimately integrates hazard and exposure to characterize risk.

- Risk mitigation involves mitigation measures to reduce or eliminate source contamination and adverse environmental impact.

- Risk management is a policy-based activity that defines risk assessment questions and endpoints to protect human health and ecological systems. It takes the scientific risk assessment and incorporates social, economic, political, and legal factors, which impinge or influence the final decision and selects regulatory actions.

B. Task Force policy decisions

Two key policy decisions were (1) an emphasis on risk mitigation early in the data gathering process and (2) less emphasis on field studies. A discussion of these two topics is in sections A and B. The Task Force also proposed specific levels of concern (LOC). These are identified in section C.

In addition, the Task Force encouraged the Office of Pesticide Programs (OPP) to consider more realistic exposure estimates when calculating estimated environmental concentrations (EECs). One recommendation was to see whether actual residue data were available within OPP that could be used to improve exposure estimates. OPP was encouraged to investigate how environmental fate data could be incorporated more effectively into exposure estimates and into the ecological risk assessment process.

The Task Force also indicated that registration and reregistration decisions could be revisited in the future. As a result, the Task Force encouraged OPP to begin development of a long-term strategy for making regulatory decisions regarding pesticide impacts and ecological risk. The strategy should aim to reduce the risk to aquatic ecosystems and should consider developing a regulatory scheme for protecting these systems from the long-term effects of pesticide use. OPP was urged to propose a research plan for improving the understanding of pesticide impacts on birds and to obtain the information needed to reduce the uncertainty in evaluating ecological risk. Finally, OPP was encouraged to consider how risk assessment and risk management could be more clearly defined and more effectively integrated, especially in the area of ecological risk.

1. Early mitigation

The Task Force concluded that risk managers must make decisions regarding ecological risk and ground-water earlier in the data gathering process. Risk managers also should rely heavily on risk mitigation when data indicate that one or more of the LOCs have been exceeded. OPP was directed to use risk mitigation to the extent feasible to improve environmental quality. To ensure that risk mitigation was successful, OPP was encouraged to consider requiring follow-up monitoring studies that would allow the Agency to evaluate the success of the imposed mitigation measures.

For ecological effects, the Task Force also emphasized that EPA "will not accept widespread and repeated mortality in the face of minor economic benefits to society. The 'widespread and repeated' standard can be met on a local or regional as well as national basis" (Ref. 4).

2. Less dependence on field studies

a. Ecological effects

In general, the Task Force concluded that (1) field studies do not provide risk managers with the kind of information that greatly enhances risk management decisions and that (2) regulatory decisions over the course of reregistration would be made in the absence of higher-tiered field studies whenever possible. As a result, OPP will no longer require avian and aquatic field studies (mesocosms), except in unusual circumstances. When field study data are lacking, decisions will be based upon other data sources, such as laboratory studies, published information and incident data. Endpoints of concern include acute and chronic toxicity. The Task Force acknowledged that indirect effects could be important, but that EPA currently does not have a testing scheme in place to accurately measure such effects within the time specified for reregistration. However, the Task Force also emphasized that if a risk manager feels that a regulatory decision cannot be made in the absence of aquatic or avian field studies, the studies can be required.

b. Ground water

The Task Force made several general recommendations with the purpose of increasing OPP's ability to make protective and timely decisions about the impact of pesticide use on ground-water quality. OPP was encouraged to make regulatory decisions where possible without requiring additional field studies, basing these decisions on the quality of existing data, environmental fate characteristics, existing monitoring databases, and modeling. By making these decisions at an earlier stage in the regulatory process, OPP will increase the likelihood that ground-water quality will not be degraded as a result of continued use of the pesticide. If existing databases do not provide an adequate basis for these regulatory decisions, risk mitigation measures may be imposed prior to the requirement for ground-water studies to confirm the impact on ground-water quality. Also, if it is unclear that the impact of mitigation measures will result in a significant improvement in protection of ground-water resources, monitoring may be required to insure its effectiveness.

C. The LOCs and Other Criteria for Regulatory Action

1. Ecological Effects

Aquatic, acute effects

- If the EEC > 1/2 LC_{50}[2], the acute aquatic risk is of high concern and may warrant regulatory action in addition to restricted use classification.
- If 1/10 $LC_{50} \leq$ EEC $\leq$ 1/2 LC_{50}, then the pesticide is considered for restricted use classification.
- If EEC < 1/10 LC_{50}, then the pesticide has a low acute aquatic risk, and no additional regulatory action is warranted.
- Incidents[5] may also trigger further assessments and will be analyzed on a case-by-base basis.

Aquatic, chronic effects

- If the EEC $\geq$ LEL[3], then the chronic aquatic risk is of high concern and may warrant regulatory action.

Avian acute effects --

- If the EEC > 1/2 LC_{50} or LD_{50}[4]/sq ft > 1/2, the acute avian risk of a pesticide is of high concern and may warrant regulatory action in addition to restricted use classification.

---

[2] The median lethal concentration necessary to effect 50% of the test population.

[3] Lowest effect level.

[4] Median lethal dose necessary to effect 50% of the test population.

- If (1) 1/5 $LC_{50} \leq EEC \leq 1/2\ LC_{50}$ or (2) $LD_{50} \geq 50$ mg/kg for granular formulations[5], then the pesticide is considered for restricted use classification.

- If (1) EEC < 1/5 LC_{50} or (2) LD_{50}/sq ft < 1/5[5], then the pesticide has low acute avian risk and no additional regulatory action will be pursued.

- Incidents[5] may also trigger further assessments and will be analyzed on a case-by-base basis.

Avian chronic, effects

- If the EEC $\geq$ LEL, then the chronic avian risk is of high concern and may warrant regulatory action.

Endangered species

The criteria that are used to determine if a pesticide may affect listed species are outlined in EPA's Standard Evaluation Procedure for Ecological Risk Assessment (Ref. 3).

2. Ground water

Mobility

- Kd[6] $\leq$ 5 mg/gm or Koc[7] $\leq$ 500 mg/gm or detection of compound 90 cm in soil profile in soil dissipation study, and

Persistence

- Soil half-life > 2 - 3 weeks, and

Risk based trigger

- The policy regarding the type of regulatory action based on human and ecological end points is still proposed and is subject to further clarification.

Detections[5]

- Detections also will play a role in the weight-of-the evidence evaluation regarding the potential of a pesticide to contaminate ground water. The nature of the detections needed to go from one regulatory action to another will be

---

[5] Not originally identified in the published findings of the Task Force.

[6] Partition coefficient.

[7] Organic carbon partition coefficient.

delineated in the proposed regulations covering restricted use classification and state management plans.

IV. Implementation Plan for Ecological Effects

The Implementation Workgroup began by focusing attention on a conceptual process to implement the Task Force policy decisions related to ecological risk. The process was designed to address two goals.

The first goal was the need to meet Congressionally-mandated deadlines for reregistration. Therefore, the process was designed to be completed in a fairly short timeframe, starting with step 4. OPP anticipates that the process generally will take less than 9 months beginning with step 4. It should be noted, however, that the process initially may take longer since it will be evolving over the course of the next year.

Timeframes have not been provided for steps 1, 2, and 3. These steps are part of OPP's routine scheduling of program activities, which will coordinate the timing of chemicals going through the implementation steps with other program activities, including reregistration.

The second goal was to make decisions that would protect environmental resources. To fulfill this goal, the process emphasizes risk mitigation and implements post-registration and reregistration monitoring as a discrete component of the process to ensure effective risk mitigation measures. The new paradigm will require OPP to identify potential mitigation measures early in the process, which is soon after an LOC exceedence is identified. Registrants are notified of the LOC exceedence and encouraged to propose risk mitigation measures during the negotiations. This results in more timely and potentially more protective decisions.

The process itself is described through a series of implementation steps, which are discussed below.

Step 1 - Define ecological risk assessment, identify endpoints of concern, and prepare preliminary usage information.

The focus of this step is to determine the kind of information and assessments that a risk manager needs to make a regulatory decision for a particular pesticide. This determination includes identifying the endpoints of concern as well as identifying the appropriate risk management questions, which may vary depending on the chemical.

This step is an important one and directly addresses one of the concerns highlighted by the Task Force, which is that the roles of the risk assessors and risk managers are not well defined or understood in the area of ecological effects. Step 1 is an attempt to address this and to more effectively integrate risk assessment and risk management.

In step 1 the use sites that will be targeted in the risk assessment also will be identified. In addition, BEAD will begin to prepare preliminary usage information on those sites. If a new chemical is being considered for registration, BEAD also will determine the need to require efficacy data.

Step 2 - Schedule preliminary risk assessment.

This is an administrative step during which a schedule is developed for the scientific review of a new pesticide or one undergoing reregistration. Enough time needs to be allocated in the schedule to allow the process to reach completion. The lead division is either SRRD or RD, depending on whether its a new pesticide or one undergoing reregistration. SRRD and RD negotiates with EFED and BEAD when developing the schedule.

Step 3 - Perform initial risk assessment.

Steps 3 and 4 essentially serve as a screen. EFED conducts the initial risk assessment that includes (1) calculating risk quotients and (2) gathering any information related to risk characterization that is readily available. The quotients will be estimated based on Kenaga (Ref. 1) and preliminary aquatic exposure scenarios.

Step 4 - Identify LOC exceedence and notify Lead Division. (Timeframe - 1 week)

Once the quotients have been calculated, they are compared to the LOC. If the LOC is not exceeded, work continues to register/reregister the pesticide. In contrast, if the LOC is exceeded, EFED notifies SRRD, RD, BEAD, and FOD of their concern.

Step 5 - Refine risk assessment and identify potential mitigation measures. (Timeframe - 6 weeks)

Once a concern has been raised by the LOC exceedence, EFED modifies the risk assessment (initial screen) by further evaluating toxicity and modifying the exposure estimates. This may be achieved by considering additional toxicity data as well as available field residue data, actual usage data, as well as utilizing more sophisticated models. Environmental fate data will be incorporated, and additional information related to risk characterization is gathered. The results are once again compared to the LOCs. If the LOCs are not exceeded, work continues to register/reregister the chemical. Depending on the certainty associated with the assessment, field monitoring may be considered.

As mentioned previously, the new paradigm emphasizes risk mitigation for ecological effects when data indicate that an LOC has been exceeded. If the LOC is still exceeded after the risk assessment has been refined, OPP will begin to identify potential mitigation measures and consult with BEAD regarding their potential feasibility.

In the past mesocosms or avian field studies were often required if the LOC was exceeded. As stated previously, the new paradigm generally places less emphasis on field studies. As a result, field studies will only be required if the circumstances are unusual, such as a new mode of action or a new chemical class. If a field study is warranted, EFED will recommend that it be required.

Step 6 - Notify registrant of LOC exceedence. (Timeframe - 3 weeks for steps 6 and 7)

This is a notification step. SRRD or RD notifies the affected registrant that OPP is concerned about the pesticide and that the registrant should consider proposing risk mitigation measures (step 8). Documentation from the refined risk assessment is included in the notification.

Step 7 - Identify cost and applicability of mitigation measures and prepare preliminary benefits assessment.

To prepare for the meeting with the registrant, OPP will evaluate the economic impact of the mitigation measures identified in step 5. To accomplish this, BEAD will develop a preliminary cost analysis of the mitigation measures. BEAD also will begin work on the preliminary benefits assessment, including identification of alternatives. In addition, OPP will develop strategy for negotiation.

Step 8 - Negotiation. (Timeframe - 4 weeks)

As indicated previously, one of the key results of the Task Force was an emphasis on risk mitigation early in the process. Under the new paradigm, SRRD or RD will negotiate with the registrants. The negotiation will consider OPP's concerns about the pesticide and will address potential risk mitigation measures. This is less than one month after the risk assessment has been refined.

In some cases, a public meeting may take place. OPP anticipates, however, that this will not occur routinely. In those cases when a public meeting is held, OPP anticipates that the 9 month timeframe may be exceeded.

In either case, the timeframe for step 9 will be determined during this step.

Step 9 - Evaluate effectiveness of proposal and consider remaining risk with respect to LOC and preliminary benefits and make a regulatory decision. (Timeframe - 4 - 12 weeks)

During step 9, the risk is re-evaluated considering the registrant's risk mitigation proposal or one that is negotiated between OPP and the registrant in step 8. A preliminary risk/benefit assessment is conducted. The tasks that need to be completed during this step include

- Evaluating the registrant or negotiated proposal and effectiveness of mitigation based on the available data,
- Characterizing the risk to the extent possible,
- Evaluating the pesticide alternatives,
- Weighing risks and benefits,
- Deciding whether to accept or reject the proposal, and
- Determining whether to require follow-up data[8] to help answer risk management questions not previously addressed and/or to monitoring the effectiveness of risk mitigation measures.

At the conclusion of this step, a decision should be made whether or not to register/reregister the pesticide based on the preliminary risk/benefit analysis. If additional data are needed, they will be required, and the results used in a later evaluation of the pesticide. These additional data can be required as a condition of registration or a confirmation of reregistration eligibility decisions. Together with changes in the benefits, they will form the analytical basis for achieving continuous improvement when revisiting the decision in the future.

If the decision is made not to register/reregister, then a full risk/benefits assessment will be performed (step 10). OPP anticipates, however, that step 9 will be the final step for most pesticides and that they will not need to continue to step 10.

Step 10 - Perform full risk/benefits assessment.

In those cases where a registration/reregistration decision cannot be made, the risk and benefit assessments that have been conducted up to this point will be expanded. EFED will need to

---

[8] Follow-up data could include aquatic or avian field studies or monitoring studies to address questions on long-term direct and indirect effects on individuals, populations, or ecosystems.

expand the alternatives assessment and fully characterize the risk based on a weight-of-the-evidence analysis, including extent of use, incident reports, species exposed, and quality of habitat. BEAD will also expand their alternatives assessment as well as conduct a full benefits assessment.

Step 11 - Make regulatory decision based on risk/benefit assessment.

Once the expanded risk/benefits assessment has been completed, a regulatory decision will be made. This requires an unreasonable risk determination and evaluating potential regulatory options. These options could include registration/reregistration, requiring additional data, label changes, and potentially cancellation. Once the options have been evaluated, SRRD or RD will make a regulatory decision. If additional data such as monitoring are needed, they will be required.

V. Implementation Plan for Ground Water

Once the plan for ecological risk was completed, the Implementation Workgroup focused their attention on developing a conceptual process to implement the Task Force policy decisions related to ground water. The results are described below. This process beginning with step 3 is also anticipated to take less than 9 months.

Similar to the plan for ecological risk, the process for ground water also was designed to fulfill the two goals -- making more timely reregistration decisions as well as more protective decisions. These goals will be achieved by focusing on risk mitigation early in the process. In the past, risk mitigation was not considered until after ground-water monitoring studies had been submitted. Under the new paradigm, risk mitigation along with potential monitoring requirements will be considered after the potential to contaminate ground water has been evaluated.

It is important to note that the Agency's policy for regulating pesticides that have the potential to contaminate ground water is still evolving. As a consequence, the steps outlined below may change or may not be followed in the same manner for all pesticides.

Step 1 - Schedule preliminary environmental exposure assessment.

This is an administrative step during which SRRD or RD develops a schedule for the science branches to review a new pesticide or a reregistration chemical. The schedule must reflect enough time to allow the process to be completed.

Step 2 - Review and evaluate environmental fate and transport data.

EFED will review and evaluate the environmental fate and transport data for the purpose of developing an environmental exposure assessment. For new pesticides and for reregistration chemicals, this effort includes evaluating laboratory and field dissipation studies. For reregistration chemicals, it also involves a cursory review of the information available in the ground-water data base.

EFED will evaluate the need for further assessment based on the weight-of-the-evidence analysis of environmental fate and transport data. If further assessment is not needed, work continues to register/reregister the pesticide.

Step 3 - Schedule ground-water evaluation and notify SRRD, RD, HED, PSPS, and BEAD. (Timeframe - 2 weeks)

If EFED determines the need for further assessment, EFED confers with other scientists to see if there is any potential ecological or human health impact from ground-water contamination. This includes conferring with the Ecological Effects Branch regarding the potential for impacts to terrestrial and aquatic species effects and the Health Effects Division regarding the potential impacts to human health.

If a concern is raised, SRRD, RD, HED, PSPS, and BEAD will be notified. EFED will work with SRRD or RD to scope out the ground-water evaluation to insure that the assessment will provide the information needed to make a regulatory decision. If further analysis is needed, it will be scheduled. BEAD will begin to prepare usage information.

Step 4 - Ground-water evaluation. (Timeframe - 1 to 4 months)

EFED conducts the ground-water evaluation as determined in step 3. EFED evaluates ground-water monitoring data and determines if a ground-water monitoring study should be required.

On the basis of this evaluation of the various kinds of data that may indicate ground-water concerns, EFED in consultation with PSPS and the other relevant Divisions will suggest risk mitigation measures, such as ground-water advisory statements or use limitations, that are appropriate to the degree and specificity of the potential contamination. In the case of data indicating that the pesticide meets or exceeds the specific criteria[9] for consideration of restricted use classification,

---

[9] These criteria will be promulgated as 40 CFR §152.170(b)(3).

EFED will consult with PSPS regarding the commencement of regulatory action (Notice of Proposed Rulemaking).

If appropriate, EFED will identify potential mitigation steps and provide a qualitative evaluation of effectiveness.

Step 5 - Notify lead division of the concern for ground-water contamination and forward usage information. (Timeframe - 1 to 2 weeks)

If a concern is raised regarding the impacts from ground-water contamination, SRRD or RD will be notified. EFED will indicate whether the chemical meets or exceeds the criteria for regulatory action and recommend potential mitigation steps. BEAD completes and forwards their usage information to SRRD, RD, and EFED.

Step 6 - Notify registrant of ground-water concern and identify the regulatory criteria that have been exceeded. (Timeframe - 4 weeks for steps 6 and 7)

This is a notification step. SRRD or RD notifies affected registrant(s) that OPP is concerned about the pesticide's potential to contaminate ground water and that OPP is considering regulatory action. Documentation from the ground-water evaluation will be sent as well. The registrant will be encouraged to propose risk mitigation measures.

Step 7 - Identify cost and applicability of mitigation measures, prepare preliminary benefits assessment, and develop strategy for negotiation.

To prepare for the negotiations, BEAD will evaluate the mitigation measures on a nationwide and local basis and develop quick cost analysis. BEAD also will perform a preliminary benefits assessment and will identify alternatives. In addition, OPP will develop strategy for negotiation and will consider potential mitigation measures and monitoring requirements.

Step 8 - Negotiation. (Timeframe - 4 weeks)

SRRD or RD will hold a meeting with the registrants and discuss potential mitigation measures and monitoring requirements. RD or SRRD also may solicit input from state and EPA Regional Offices, and in some cases may hold a public meeting. OPP anticipates, however, that this will not occur routinely. In those cases when a public meeting is held, OPP anticipates that the 9 month timeframe may be exceeded.

In either case, the timeframe for step 9 will be determined during this step.

Step 9 - Evaluate final registrant proposal and make risk/benefit determination with respect to ground-water contamination. (Timeframe - 4 to 12 weeks)

During the final step in the process, the registrant proposal is evaluated. Mitigation, monitoring data, and the relative potential of the alternatives to contaminate ground water are considered. This step also includes weighing risks and benefits and determining whether to accept or reject the registrant proposal. This step, as well as the process, ends by making a regulatory decision and by requiring monitoring data, if needed.

V. Clarification of Task Force Results

Since the new paradigm was developed, several general questions arose that warranted clarification. These questions are described in this section along with a response.

A. Ecological effects

1. Applicability of new paradigm to registration and special review.

One question that arose was the applicability of the new paradigm to the registration program and to special reviews. The policy was clear in its focus on reregistration. In contrast, it was not clear how registration and special review were considered during the course of developing the new paradigm. This is of concern because registration, reregistration, and special reviews are different regulatory processes. Reregistration operates under Congressionally-mandated deadlines, while registration and special reviews do not. Also, the information that may be available for a pesticide that has been in use for some time may be quite different than that for a new pesticide being considered for registration. For example, incident data, which was highlighted to be an important part of risk characterization in the new paradigm, may be available for old pesticides, but not for new.

Concern also was raised because of the de-emphasis on field studies, and the impact on ecological risk assessment. Since field studies will generally not be required to determine ecological effects, pesticides could be registered without any field data. Therefore, certain routes of exposure such as dermal and inhalation, indirect effects such as reduced food availability, and ecosystem effects will not be considered.

Response: It is important to revisit the overriding emphasis of the policy. That is, risk managers should only ask for a study when the information from such a study will improve our ability to make decisions. Applied to the registration program, this means that if OPP has significant uncertainty about the effects of a new chemical in the field which cannot be resolved with out a field study, then a field study should be considered. However, before requiring such a study (or any other study) OPP should know what actions to take. This relates to knowing how to use the results of a study before requiring it.

2. Regulatory decisions in the absence of avian field studies.

A question also was raised regarding the statement "The AA agrees with workgroup conclusion that the avian field study provides very limited new information to an avian risk assessment, and that such field studies confirm the results of the lab studies" (Ref. 4). Commenters have asked to be shown the scientific basis for this statement.

Concern also was raised regarding the conditions under which a field study would be required. The Task Force indicated that field studies could be required under unusual circumstances, yet did not make it clear what these "unusual circumstances" would be.

Response: The Task Force conclusion that the avian field study provides very limited new information to an avian risk assessment and thus would only be required under unusual circumstances was based to a large degree upon two factors. First, a consensus of Task Force members indicated that the results of the avian field studies reviewed by OPP always confirmed predictions of adverse effects, primarily bird mortality, based upon lower tiered studies. Second, a preliminary analysis of "in-house" avian field studies supported this conclusion. The Task Force recognized that the conclusion was based upon limited data on organophosphate and carbamate pesticides. Thus, the "unusual circumstances" where avian field studies may be required in the future could include new pesticides which exceed the LOCs for birds and which have chemistry and mode of actions significantly different from organophosphate and carbamate pesticides. However, other cases may arise which also may warrant the requirement of a field study.

3. New LOCs were identified.

Commenters also indicated that two new LOCs are identified by the new paradigm. This was questioned because scientific justification was not provided and the LOCs have not undergone adequate analysis. Specifically, the LOCs are (1) LD_{50}/sq ft > 1/2 and (2) EEC > 1/2 LC_{50}.

Response: The new LOCs are based on the results of a preliminary, retrospective analysis of 20 field studies. The results support the conclusion that bird kills in the field may occur at these LOCs.

Also, the LOCs for ecological effects have been established based on Agency regulations and guidance[10]. In 1975, a Rebuttable Presumption Against Registration (RPAR, the predecessor to Special Review), could be initiated when the estimated environmental exposure was equal to or greater than one-half the acute toxicity value for aquatic organisms. The "trigger" for birds at that time was equal to the acute avian toxicity value with no "safety factor". Internally, OPP always believed that some safety factor was needed, but the majority of pesticides for which avian field studies were required had estimated exposures much greater than the acute toxicity values for birds. Thus, OPP never established a number that captured the notion of "approaching" the bird toxicity value.

In the recent proposed revision to Part 158, OPP stated that the avian field study would be required when the estimated environmental exposure is equal to or greater than one-half the acute avian toxicity value. This is identical to the trigger for the aquatic mesocosm.

4. Other LOCs.

A question also was raised regarding endangered species and mammals, whether they are endangered or not. The new paradigm did not specifically address the use of the existing LOCs for endangered species and mammals or identify new LOCs. It is not clear how these endpoints of concern will be addressed by the new paradigm.

Response: The LOCs for endangered species (Ref.3) are unaffected by the new paradigm. While not specifically addressed in the findings of the Task Force (Ref. 4), the LOCs for wild mammals will be identical to those for birds.

B. Ground water

1. Ability to meet the protective and timely goals of the new paradigm.

A question was raised whether or not making decisions more quickly could lead to less protective decisions for ground water.

---

[10] Examples include the regulations for Registration, Reregistration and Classification (40 FR 28242); the current regulations (40 CFR 154.7); the criteria for initiation of Special Review (40 CFR 154); Criteria for Restriction for Use by Certified Applicators [40 CFR 152.170(c)]; 1988 Guidance Document for Conducting Terrestrial Field Studies (Ref. 3)

Although the intent of the new paradigm is to make more protective and timely decisions, it may be difficult because or the uncertainty associated with ground-water monitoring data and computer modeling results if field data from appropriately designed and conducted field studies are not available.

Response: The Task Force emphasizes that OPP needs to make decisions that "result in a significant and prompt improvement in environmental quality. More decisions--environmentally protective decisions, with a major focus on remediation, where appropriate-- will be made with far greater speed under the new paradigm [...]". If the existing database is inadequate to support decisions that would be protective, and that database can be improved by requirement of a ground-water field study, the decision to require such a field study should be made at an earlier stage in the registration and reregistration processes. Also, if the effectiveness of specific remediation options cannot be confirmed, monitoring should be required to ensure that the Agency is acting in a more protective mode by requiring them.

The Task Force recognized the value of the information provided by the ground-water field studies. These studies are designed to gather data to support registration and reregistration decisions. If the alternative to requiring a ground-water field study is that the Agency is compelled to make decisions that are less protective, even though these decisions are made more rapidly, this was clearly not the intended outcome of the new paradigm.

As a consequence, it is essential for OPP to closely monitor the implementation of the new paradigm. First, OPP will track the process to ensure that the timeframes and steps outlined in sections IV and V are followed. Second, it will include evaluating the results of the process. If the decisions that result are not made in a timely manner and are not protective, then OPP will need to step back and re-evaluate the implementation plan. This will be true for ground-water as well as for ecological effects.

VI. Conclusions and Long-term Research and risk Assessment Improvement Goals

OPP believes that the implementation plan described in this paper will help to improve OPP's ability to meet the reregistration deadlines that have been established and to evaluate registration actions in a more timely manner. The resulting decisions will achieve the Task Force goals by being environmentally protective and more timely. OPP believes it is taking a step in the direction of environmental protection by its emphasis on early decision-making and risk mitigation.

To provide continued scientific support for the new paradigm, OPP has identified several areas which could benefit from additional research. In some cases, work has already begun.

A. Ecological effects

The following project has already begun and a report should be completed by spring of 1994:

- Retrospective Analysis of Terrestrial Field Studies

  A synopsis of all "in-house" terrestrial field studies will be done by the EPA's Office of Research and Development"s Environmental Research Laboratory in Corvalis, Oregon. The results will be summarized and analyzed to verify the preliminary analysis of 20 field studies and to determine if terrestrial field studies validate OPP's presumption of terrestrial risk.

Additional projects are planned to begin in October 1993, and reports should be available by the end of 1994. Some projects may continue beyond 1994 as the need for additional work becomes apparent:

- Retrospective Analysis of Aquatic Field Studies

  A synopsis of all completed mesocosms and pond studies will be done by ORD's Environmental Research Laboratory at Duluth, Minnesota. The results will be summarized and analyzed to determine if they validate the presumption of aquatic risk.

- Risk Mitigation and Monitoring Guidance Projects

  In response to the Task Force emphasis on risk mitigation, this project will identify and advance risk reduction and mitigation measures for terrestrial organisms and their habitats threatened by pesticide use.

  In addition, terrestrial effects monitoring guidance will be developed as a tool to measure the effectiveness of risk mitigation measures. Guidance should include residue and effects on terrestrial non-target organisms. The guidance may be tiered and should provide site-specific guidance.

  Two other projects will look into risk reduction and mitigation measures as well. One project will focus on aquatic organisms and their habitats threatened by pesticide use and the other on nontarget plants. Monitoring guidance also will be developed.

- **Wildlife Utilization Project**

  The Task Force emphasized the development of additional tools to further characterize the risk of pesticide use to wildlife, including the exposure of wildlife to pesticide residues. This project provides for the development of a computerized (e.g., dBase III+) database using Gusey and Maturgo (1973) wildlife utilization data. Data would be used by scientists to better characterize terrestrial exposure and risk. The data would serve as a starting point for the development of a wildlife utilization model. Additional literature and survey information would eventually be incorporated into the data base.

- **Evaluation of Aquatic Habitat Project**

  The Task Force emphasized the need for additional tools to characterize the exposure and risk of pesticides to aquatic organisms. These tools may be available from groups in other parts of the Agency or outside the Agency.

  This project will coordinate the search for existing databases and maps for characterizing aquatic organism exposure and risk from pesticides use. High value use areas for aquatic organisms, such as areas adjacent to productive streams/rivers, marshes, as well as critical habitats for endangered and threatened species, will be identified.

  This project also involves participation on Agency Habitat and Biodiversity workgroups.

- **Evaluation of Terrestrial Habitat Project**

  This project will coordinate the search for existing databases and maps for characterizing wildlife exposure and risk from pesticides use. High value use areas for wildlife, such as areas adjacent to wildlife refuges, bird migratory flyways, critical habitat for endangered and threatened species, will be identified.

- **Incident Data Project**

  This project is a continuation and enhancement of the "Ecological Incident Information System" and involves entering incident data on pesticides and other toxic chemicals. Data sources include OPP files, the states, and other federal agencies such as the U.S. Fish and Wildlife Service. Currently, 200 incident records have been evaluated and entered.

  The goal of this project is to provide the Agency scientists, registrants, the Regions, the states, and outside scientists with a file structure so that ecological effects data can be easily accessed and used to characterize ecological risk assessments.

- **EEB One-Liners/Chesapeake Bay Project**

  This project is a continuation and enhancement of current computerized Ecological Effects Branch's (EEB) "One-liner" files and includes all pesticides for which eco-toxicity exists in EEB files. Currently, EEB has approximately 4400 records for 209 chemicals. Additional data files need to be entered, and the data base needs to be updated and verified.

  The goal of this project is to provide the Chesapeake Bay Project, OPP scientists, registrants, the Regions, the states, and outside scientists with a file structure so that ecological effects data can be easily accessed and used for ecological risk assessments.

- **Value of [Additional] Information Project**

  Decisions to ask for higher tiered data to support registration of pesticides is clearly a risk management decision. Risk managers need to know what additional information higher tiered data will provide for risk assessment, and what value it will add to risk management decisions.

  Scientific rationale will be developed supporting decisions to require or not require tier 2, 3 and 4 data. Factors considered in the rationale will include: Tier 1 toxicity data, fate and transport characteristics, use pattern, LOCs, etc.

- **Species Tested versus Species Exposed Project**

  The Task Force considered an analysis of the species exposed versus the species tested as an essential piece of information in making regulatory decisions for ecological effects. Existing eco-toxicity databases will be identified and collected, and databases will be analyzed to compare the sensitivity of various species tested to pesticides. Lists of species identified (and likely exposed to pesticides) with different use patterns will be developed. Comparisons of species tested and species exposed will be made. Extrapolations from one species to another based on body weight and other factors will be considered.

- Eco-Effects Modelling Project

  A number of models exist or are currently being developed. In order to determine which, if any of these models will assist in characterizing ecological risk, this project will provide a review and evaluation of eco-effects computer models such as FGETS, RAMAS, PIRANHA and LERAM, etc. Hopefully, one or more of these models will be useful to predict and/or characterize adverse effects in non-target organisms, their populations and communities.

  The endpoints to be considered will include direct and indirect mortality, reproductive/chronic effects, bioaccumulation, age-structured population dynamic changes and bioenergetic ecosystem effects.

- Uncertainty in Ecological Risk Characterization Project

  Uncertainty exists in every step of ecological risk characterization, which includes identification of endpoints of concern, effects and exposure characterization, quotient calculations, refinement of quotient estimates, modelling, etc. Identifying the uncertainty associated with this information will place it in perspective and may lead to better regulatory decisions.

  Existing eco-toxicity, exposure, and risk data bases will be analyzed for variability. Best approaches and statistical techniques will be selected with statistical assistance. Based upon these analyses and existing analyses from the literature, statements of uncertainty will be developed for use in ecological risk assessments.

B. Ground water

- Tools must be developed and refined to enhance OPP's modeling and GIS capabilities. In particular, model validation and enhancement of pesticide leaching and runoff models will greatly improve OPP's ability to rapidly respond to questions about the impact of new uses of pesticides or continued use of pesticides on ground and surface water quality. Currently, there are uncertainties about the predictive value of these models. Significantly more research and field work is needed before these questions are resolved.

- The emphasis of the new paradigm on risk mitigation is very clear. What remains to be established is the overall impact of individual risk mitigation measures on environmental quality. These effects should be quantified by requirement of monitoring to (1) determine the impact of each measure on changing the quality of ground water, surface water, air, and the ecosystem, and (2) to establish an integrated approach to resource assessment to ensure that benefits to one environmental medium that result from individual mitigation measures are not offset by significant costs incurred by others.

VII. References

1. Kenaga, E.E. 1973. Factors to be Considered in the Evaluation of the Toxicity of Pesticides to Birds in Their Environment. Environmental Quality and Safety. Academic Press, N.Y. II:166-181.

2. U.S. Environmental Protection Agency. 1986. Standard Evaluation Procedure for Ecological Risk Assessment. EPA 540-9-86-167.

3. U.S. Environmental Protection Agency. 1988. Guidance Document for Conducting Terrestrial Field studies. EPA 540/09-88-109.

4. U.S. Environmental Protection Agency. 1992. Memorandum from Linda J. Fisher, Assistant Administrator, to Douglas Campt, Director, Office of Pesticide Programs, dated October 29, 1992. SUBJECT: Decisions on the Ecological, Fate, and Effects Task Force.

GUIDELINES, STANDARD EVALUATION PROCEDURES, AND TECHNICAL GUIDANCE DOCUMENTS USED IN PESTICIDE REGISTRATION AND REREGISTRATION

October, 1974: American Institute of Biological Sciences, under contract with USEPA (Contract No. 68-01-2457) completes the Draft Report Document ANALYSIS OF SPECIALIZED PESTICIDE PROBLEMS, VOLUME VI, WILDLIFE TOXICOLOGY STUDY. This document provides guidance concerning the wildlife toxicology studies to perform and recommended tiered avian and wild mammal toxicity tests.

40 CFR No. 123, June 25, 1975: USEPA published Guidelines for Registering Pesticides in United States. Provides guidance on what studies to do and how to do them.

40 CFR No. 132, July 10,1978: USEPA published REGISTRATION OF PESTICIDES IN THE UNITED STATES Proposed Guidelines. Provides further guidance on what studies to do and how.

August, 1978: American Institute of Biological Sciences, under contract with USEPA (Contract No. 68-01-2457) finalizes the document CRITERIA AND RATIONALE FOR DECISION MAKING IN AQUATIC HAZARD EVALUATION. This document provides guidance about what aquatic studies to perform and USEPA's tiered aquatic toxicity tests.

40 CFR Part 158.145, 158.150, and 158.155, October 24, 1984: USEPA published Data Requirements for Pesticide Registration; Final Rule. This publication provides guidance on what wildlife, aquatic, plant and beneficial insect studies to perform to support various use patterns.

June, 1985: USEPA begins publishing Standard Evaluation Procedures (SEPs) for each study requirement outlined in the Agency's guidelines. Each SEP details what makes an acceptable guideline study (SEE TABLE 1).

June, 1986: USEPA published its Ecological Risk Assessment SEP which outlines how required guideline studies are used in the Agency's ecological risk characterization process.

March, 1988: USEPA published TECHNICAL GUIDANCE DOCUMENT AQUATIC MESOCOSM TESTS TO SUPPORT PESTICIDE REGISTRATIONS. This paper describes how to perform an aquatic mesocosm test for assessing aquatic toxicity (SEE TABLE 1).

September, 1988: USEPA published Guidance Document for Conducting Terrestrial Field Studies. This document details how to perform terrestrial field studies to determine terrestrial toxicity (SEE TABLE 1).

Phase III Guidance Package. (12/21/89). This package, sent to all pesticide registrants, contained acceptance criteria check list for all EEB guidelines requirements, format for summarizing each of the EEB guideline requirements, and a sample of the acceptance criteria and format for the fish early life-stage test.

Standard Evaluation Procedures	NTIS Number	EPA Number
Subdivision E, Haz Eval: Wildlife & Aquatic Organisms	PB83-153908	540/09-82-024
Subdivision E, HE: Wildlife & Aquatic Organisms-Addendum 1	PB86-248176	540/09-86-152
Subdivision E, HE: Wildlife & Aquatic Organisms-Addendum 2	PB87-207700	540/09-87-198
Subdivision E, HE: Wildlife & Aquatic Organisms-Addendum 3	PB88-117288	540/09-88-006
Subdivision J, Haz Eval: Nontarget Plants	PB83-153940	540/09-82-020
Subdivision L, Haz Eval: Nontarget Insects	PB83-153957	540/09-82-019
Subdivision L, Haz Eval: Nontarget Insects- Addendum l	PB88-117296	540/09-88-005
Subdivision N, Chemistry: Environmental Fate - Addendum 4	PB88-159900	540/09-88-021
Subdivision N, Chemistry: Environmental Fate - Addendum 5	PB88-161187	540/09-88-025
Subdivision N, Chemistry: Environmental Fate - Addendum 6	PB88-161195	540/09-88-026
Subdivision N, Chemistry: Environmental Fate - Addendum 7	PB88-191721	540/09-88-050
Subdivision N, Chemistry: Environmental Fate - Addendum 8	PB88-191739	540/09-88-051
Acute Tox Test for Est/Marine Org (Mollusc 48HR Embryo Larv)	PB86-129335	540/09-85-012
Acute Tox Test for Est/Marine Org (Mollusc 96HR Flow-Thru Sh)	PB86-129327	540/09-85-011
Acute Tox Test for Est Marine Org (Est Fish 96-HR)	PB86-129301	540/09-85-009

Acute Tox Test for Freshwater Fish	PB86-129177	540/09-85-006
Acute Tox Test for Freshwater Invertebrates	PB86-129169	540/09-85-005
Acute Tox Test: Est Marine Organisms (Shrimp 96-HR)	PB86-129319	540/09-85-010
Avian Dietary LC_{50} Test	PB86-129293	540/09-85-008
Avian Reproduction Test	PB87-209177	540/09-86-139
Avian Single-Dose Oral LD_{50} Test	PB86-129285	540/09-85-007
Daphnia Magna Life-Cycle (21-day REN) Chronic Toxicity Test	PB87-209730	540/09-86-141
Ecological Risk Assessment	PB86-247657	540/09-86-167
Field Testing for Pollinators	PB87-209714	540/09-86-140
Fish Early-life Stage Test	PB87-209722	540/09-86-138
Fish Life-cycle Toxicity Test	PB87-209169	540/09-86-137
Honey Bee - Acute Contact LD_{50}	PB86-154580	540/09-85-002
Honey Bee - Tox of Residues on Foilage	PB86-129244	540/09-85-003
Non-Target plants: Aquatic Field Testing - Tier 3	PB87-101705	540/09-86-136
Non-Target plants: Growth and Repro of Aquatic Plants-Tiers 1/2	PB87-102232	540/09-86-134
Non-Target plants: Seed Germ/Seedling Emergence, Tiers 1 & 2	PB87-102240	540/09-86-132
Non-Target Plants: Target Area Testing	PB87-101689	540/09-86-130
Non-Target Plants: Terrestrial Field Testing - Tier 3	PB87-101697	540/09-86-135
Non-Target Plants: Vegetative	PB87-102224	540/09-86-133
Wild Mammal Toxicity Test	PB86-129251	540/09-85-004
Terrestrial Field Studies	PB89-124580	540/09-88-109
Aquatic Mesocosm Tests	PB88-172465	540/09-88-035

(Entries are alphabetical by chapter of first occurrence.)

Executive Summary

Aquatic Risk Assessment and Mitigation Dialogue Group (Aquatic Dialogue Group)

Because of the importance of the new paradigm in pesticide regulatory decisions, it was felt that an open scientific dialogue regarding aquatic risk assessment and mitigation options would assist in the technical implementation of the new paradigm. The group, composed of representatives from USEPA (regulators and researchers), agrichemical companies, academia, and environmental and agricultural interest groups, was formed under joint sponsorship by the American Crop Protection Association and USEPA.

Ecological Risk Assessment

A process that evaluates the likelihood that adverse ecological effects may occur or are occurring as a result of exposure to one or more stressors. A risk does not exist unless

- the stressor has the inherent ability to cause one or more adverse effects AND
- it co-occurs with or contacts an ecological component (*i.e.*, organisms, populations, communities or ecosystems) long enough and at a sufficient intensity to elicit the identified adverse effect.

USEPA has defined that the process is divided into three phases:

1. Problem formulation—What are the scientific and regulatory needs and the scope of the issue?
2. Risk analysis—characterization of exposure to predict or measure the spatial and temporal distribution of a stressor and its co-occurrence or contact with the ecological components of concern
3. Risk characterization—using the results of the exposure and effects analyses to evaluate the likelihood of adverse ecological effects associated with exposure to the stressor

The assessment should include a summary of the assumptions used, the scientific uncertainties and the strengths and weaknesses of the analyses. In addition, the ecological significance of the risk is discussed with consideration of the types and magnitudes of the effects, their spatial and temporal patterns, and the likelihood of recovery.

EEC

Estimated Environmental Concentration; an estimate of the pesticide exposure that non-target organism might experience as a result of off-target movement or direct application of a pesticide. There is a loosely defined tiered system for obtaining increasingly refined EEC values; this proceeds from first approximation, worst-case simple calculations to highly sophisticated probabilistic modeling.

The EEC is often referred to as a single value, but in reality a probability distribution should be used to describe possible EEC values.

EMWG
Exposure Modeling Work Group, functioning under FIFRA

FIFRA
Federal Insecticide, Fungicide and Rodenticide Act. Designed to protect against "any unreasonable risk to man or the environment, taking into account the economic, social, and environmental costs and benefits of the use of any pesticide"

New paradigm
A phrase now used to refer to USEPA's "New Ecological Risk paradigm" as outlined in a letter from Linda Fisher (Assistant Administrator, USEPA) dated October 29, 1992. This is the driving force for the Aquatic Risk Assessment and Mitigation Dialogue Group.

OPP
USEPA's Office of Pesticide Programs (now the Office of Pollution Prevention and Toxic Substances [OPPTS])

OPPTS
USEPA's Office of Pollution Prevention and Toxic Substances (formerly the Office of Pesticide Programs [OPP])

SDTF
Spray Drift Task Force. A joint project of 29 pesticide companies that was initiated in 1990. Its objective is to produce a generic spray drift data base (and, if possible, a predictive model) for pesticide registration and reregistration decisions that affect the member companies. The project is expected to take 4 to 5 years with results available in 1995.

USEPA
United States Environmental Protection Agency

Introduction

BEAD
Biological Evaluation and Analysis Division of the OPP, one of three (also HED and EFED) scientific divisions responsible for scientific data evaluation and decision making on agricultural benefit assessments, reviewing data from pesticide registrants and other sources to develop exposure, hazard, or economic profiles and risk assessments

EFED
Environmental Fate and Effects Division of the OPP, one of three (also HED and BEAD) scientific divisions responsible for scientific data evaluation and decision making on environmental risk assessments, reviewing data from pesticide registrants and other

sources to develop exposure, hazard, or economic profiles and risk assessments. EFED encompasses two branches (EFGWB and EEB) intimately concerned with the deliberations of the Aquatic Dialogue Group.

HED
Health Effects Division of the OPP, one of three (also EFED and BEAD) scientific divisions responsible for data evaluation and decision making on human health risk assessments, reviewing data from pesticide registrants and other sources to develop exposure, hazard, or economic profiles and risk assessments

LOC
Level of concern; in OPP regulatory decision making, LOCs consist of quotient comparisons of estimated hazard (acute and chronic toxicity for fish, invertebrates, and plants) with the EEC. In addition, values may be derived for endangered species and supplemented with field data or incident data for the risk characterization.

RED
Reregistration Eligibility Decisions

RD
Registration Division of the OPP, one of two (also SRRD) divisions responsible for risk management data evaluation and decision making

SRRD
Special Review and Reregistration Division of the OPP, one of two (also RD) divisions responsible for risk management data evaluation and decision making

Standard Evaluation Procedure
Internal USEPA document describing key issues that a USEPA reviewer (or an industry study reporter) must be aware of to properly interpret a required study. An SEP for Ecological Risk Assessment was published by USEPA in 1986.

Chapter 1

Monitoring Studies
In this context, monitoring studies are thought of as field programs designed to provide an estimate of parameters (*e.g.*, aquatic residues, sediment residues, or some biological criterion) across major portions of the use area of a product. The scope of such studies is regional.

NOEC
No-Observed-Effect Concentration

Chapter 2

AEC
Actual Environmental Concentration

EWOFFT

European Workshop on Freshwater Field Tests, Potsdam, Germany, June 25-26, 1992

GIS

Geographical Information Systems

GLP

Good Laboratory Practice

GLWQG

Great Lakes Water Quality Guidelines; proposed testing requirements that are generally consistent with the data requirements for USEPA water quality criteria

LOD

Level Of Detection

LOEC

Lowest-Observed-Effect Concentration

MATC

Maximum Acceptable Toxicant Concentration

Mesocosm

A series of at least 12 large (>400 m^3 & 0.1 acre) outdoor manmade ponds designed to meet the requirements of the USEPA mesocosm guidance document. The ponds are treated with test chemical, and effects on plants, invertebrates, and fish are measured. The guiding principle for stocking and construction is that the replicated ponds should be as similar as possible.

Microcosm

Multiple meanings, but for Aquatic Dialogue Group purposes, refers to outdoor test systems significantly smaller than mesocosms

NRC

National Research Council

Small-scale Runoff Studies

Field runoff investigation generally employing replicated small plots (0.1 to 0.25 acres) of even slope which are irrigated with rainfall simulators designed to mimic natural rainfall impact energies AND apply water evenly and at constant intensity for a predetermined period. They are designed to provide information for calibrating simulation models which can then be used to predict runoff from large watersheds under the full range of possible natural rainfall conditions.

Tier 1 Toxicity Tests

Acute toxicity studies on warm-water fish (bluegill), cold-water fish (rainbow trout), and aquatic invertebrates (Daphnia) [occasionally, also estuarine fish, shrimp, and mollusc and honey bees]. The endpoint is LC50 data, and results frequently "trigger" Tier 2 studies. Studies are conducted in clean water conditions.

Tier 2 Toxicity Tests

To determine a NOEC and/or LOEC for serious effects such as reduced survival, growth, and reproduction. Fish early life-stage studies using rainbow trout, brook trout, or fathead minnows as well as an aquatic invertebrate chronic toxicity study using Daphnia. Studies are conducted in clean water conditions. Where the EEC is greater than the NOEC or LOEC, unreasonable effects are presumed.

Tier 3 Toxicity Tests

Full life-cycle studies to determine the NOEC and/or LOEC; aquatic organism accumulation and food chain transfer studies may also be required. Triggered if EEC > 0.1 NOEC; a hazard to aquatic organisms is assumed if the EEC exceeds the adjusted NOEC.

Tier 4 Toxicity Tests

Field or mesocosm testing is the final step in the tiered process.

Chapter 3

AGDISP

An aerial spray drift model, a derivative for crop uses of the Forest Service Spray Drift Model

DESDM

DowElanco Spray Drift Model, an aerial spray drift model

EPIC

Erosion Productivity Impact Calculator

EXAMS

Exposure Analysis Modeling System, a primary surface water model

FARM

Field Agricultural Monitoring Manual, from USEPA

GLEAMS

Groundwater Loading Effects of Agricultural Management Systems, a primary runoff/soil erosion model

HSPF

Hydrological Simulation Program Fortran

PRZM2

Pesticide Root Zone Model, a primary runoff/soil erosion model

SWRRB
Simulator for Water Resources in Rural Basins

WASP5
Water Quality Analysis Simulation Program, a primary surface water model

Chapter 4

BMP
Best Management Practices

IPM
Integrated Pest Management

LEPA
Low-Energy Precision Application

PPI
Pre-Plant Incorporated

SCS
Soil Conservation Service

ULV
Ultra-Low Volume application

Chapter 5

ASCS
Agricultural Stabilization and Conservation Service

CAMPS
Computer Assisted Management Program System

EMAP
Environmental Monitoring and Assessment Program

FOCS
Field Office Computing System

FSA
Food Security Act

RBP
Rapid Bioassessment Protocols

USDA
U.S. Department of Agriculture

Appendix B

EEB

USEPA OPP's Ecological Effects Branch. This group is responsible for evaluating the effect of pesticides in the environment and evaluates the significance of the potential exposure levels estimated by Environmental Fate and Groundwater Branch [EFGWB].

Other

MEC

Measured Environmental Concentration; a measured concentration from a typical site under real-world conditions. MECs are "hard" numbers but represent only the value corresponding to one particular combination of environmental variables and timing.

Primary Exposure Models

USEPA EFGWB's Surface Water Section has identified certain simulation models that registrants should use if an aquatic exposure question needs addressing. These models are currently PRZM2 or GLEAMS for leaching and runoff modeling with EXAMS or WASP5 used to assess the concentration to be expected when chemicals enter water bodies.

Probabilistic Exposure Models

Using a standard simulation model (or suite of models), it is possible to run large numbers of iterations using progressive variation of the expected ranges of key parameters (*e.g.*, weather factors, ranges of soil properties, ranges of test chemical properties) to produce a distribution of exposure estimates reflecting all the possible coincidences of the variables that may affect environmental exposure. This distribution can then be analyzed to gain an understanding of the likelihood of exposure exceeding any particular value.

Restricted Use Product

One available for use only by or under the supervision of an applicator certified to have completed pesticide safety training

SAP

Scientific Advisory Panel established under the authority of FIFRA to help USEPA ensure the effectiveness and quality of the scientific analyses that support its regulatory decision

Secondary Exposure Models

In certain cases, the primary simulation models do not meet all the scientific questions for understanding potential exposure for a particular product or use pattern. In these cases, additional simulation models may prove effective, these secondary models can be run at the discretion of the registrant provided that primary model data is also submitted and that the reason for the use of the "additional" model is justified.

Tiered Testing

An efficient method of arranging tests so that the more complicated tests are performed less often and only when the simpler tests indicate the need for further testing.

RECOMMENDED READING

Improving aquatic risk assessment under FIFRA. Report of the Aquatic Effects Dialog Group. 1992. RESOLVE, World Wildlife Fund.

Touart, S.W. 1988. Aquatic mesocosm tests to support pesticide registration. Hazard Evaluation Division technical guidance document. EPA 540/09-88-035. Office of Pesticide Programs, USEPA, Washington, DC.

Urban, D.J. and J.N. Cook. Ecological Risk Assessment. 1986. Hazard Evaluation Division Standard Evaluation Procedure. EPA-54019-83-001. Office of Pesticide Programs, USEPA, Washington, DC.

USEPA. 1992. *Framework for Ecological Risk Assessment.* EPA/630/R-92/001. Washington, DC.